Ensemble Forecasting Applied to Power Systems

Ensemble Forecasting Applied to Power Systems

Special Issue Editors

Antonio Bracale
Pasquale De Falco

MDPI • Basel • Beijing • Wuhan • Barcelona • Belgrade • Manchester • Tokyo • Cluj • Tianjin

Special Issue Editors
Antonio Bracale
University of Naples Parthenope
Italy

Pasquale De Falco
University of Napoli Parthenope
Italy

Editorial Office
MDPI
St. Alban-Anlage 66
4052 Basel, Switzerland

This is a reprint of articles from the Special Issue published online in the open access journal *Energies* (ISSN 1996-1073) (available at: https://www.mdpi.com/journal/energies/special_issues/Ensemble_Forecasting_Power_Systems).

For citation purposes, cite each article independently as indicated on the article page online and as indicated below:

LastName, A.A.; LastName, B.B.; LastName, C.C. Article Title. *Journal Name* **Year**, *Article Number*, Page Range.

ISBN 978-3-03928-312-5 (Pbk)
ISBN 978-3-03928-313-2 (PDF)

© 2020 by the authors. Articles in this book are Open Access and distributed under the Creative Commons Attribution (CC BY) license, which allows users to download, copy and build upon published articles, as long as the author and publisher are properly credited, which ensures maximum dissemination and a wider impact of our publications.

The book as a whole is distributed by MDPI under the terms and conditions of the Creative Commons license CC BY-NC-ND.

Contents

About the Special Issue Editors . . . . **vii**

Preface to "Ensemble Forecasting Applied to Power Systems" . . . . **ix**

Antonio Bracale, Guido Carpinelli and Pasquale De Falco
Developing and Comparing Different Strategies for Combining Probabilistic Photovoltaic Power Forecasts in an Ensemble Method
Reprinted from: *Energies* **2019**, *12*, 1011, doi:10.3390/en12061011 . . . . **1**

Musaed Alhussein, Syed Irtaza Haider and Khursheed Aurangzeb
Microgrid-Level Energy Management Approach Based on Short-Term Forecasting of Wind Speed and Solar Irradiance
Reprinted from: *Energies* **2019**, *12*, 1487, doi:10.3390/en12081487 . . . . **17**

Masoud Sobhani, Allison Campbell, Saurabh Sangamwar, Changlin Li and Tao Hong
Combining Weather Stations for Electric Load Forecasting
Reprinted from: *Energies* **2019**, *12*, 1510, doi:10.3390/en12081510 . . . . **45**

Tomasz Serafin, Bartosz Uniejewski and Rafał Weron
Averaging Predictive Distributions Across Calibration Windows for Day-Ahead Electricity Price Forecasting
Reprinted from: *Energies* **2019**, *12*, 2561, doi:10.3390/en12132561 . . . . **57**

Peng Li, Chen Zhang and Huan Long
Solar Power Interval Prediction via Lower and Upper Bound Estimation with a New Model Initialization Approach
Reprinted from: *Energies* **2019**, *12*, 4146, doi:10.3390/en12214146 . . . . **69**

Mohamed Lotfi, Mohammad Javadi, Gerardo J. Osório, Cláudio Monteiro and João P. S. Catalão
A Novel Ensemble Algorithm for Solar Power Forecasting Based on Kernel Density Estimation
Reprinted from: *Energies* **2020**, *13*, 216, doi:10.3390/en13010216 . . . . **87**

John Boland
Characterising Seasonality of Solar Radiation and Solar Farm Output
Reprinted from: *Energies* **2020**, *13*, 471, doi:10.3390/en13020471 . . . . **107**

About the Special Issue Editors

Antonio Bracale (Associate Professor) received a Ph.D. degree in Electrical Energy Conversion from the Second University of Napoli (currently University of Campania Vanvitelli), Italy in 2005. He currently is with the Department of Engineering, University of Napoli Parthenope, Italy. His research interests mainly focus on power quality, power system analysis, and energy forecasting. He serves as the Editor of the journal *International Transactions on Electrical Energy Systems* and he was Guest Editor of two Special Issues in MDPI journals.

Pasquale De Falco (Post-doc researcher) received a Ph.D. degree in Information Technology and Electrical Engineering from the University of Napoli Federico II, Italy in 2018. He currently is with the Department of Engineering, University of Napoli Parthenope, Italy. His research interests mainly focus on energy forecasting, energy data analytics, and probabilistic methods applied to power systems. He serves as Editor of two MDPI journals (*Electronics* and *Forecasting*) and he was Guest Editor of three Special Issues in MDPI journals.

Preface to "Ensemble Forecasting Applied to Power Systems"

Modern power systems are affected by many sources of uncertainty, driven by the spread of renewable generation, by the development of liberalized energy market systems and by the intrinsic random behavior of the final energy customers. Forecasting is, therefore, a crucial task in planning and managing modern power systems at any level, from transmission to distribution networks, and in the new context of smart grids. Many important operations are scheduled and performed on the basis of predictions of several energy-related variables, such as non-controllable generation, loads, energy prices, and power quality indicators. Forecasts of these variables at different lead times, ranging from several minutes to several days, are needed by the operators in order to suit different applications and scenarios.

The application of forecasting techniques to power systems, in both deterministic and probabilistic frameworks, is a topic that is still far from being fully explored. Recent trends suggest the suitability of ensemble approaches in order to increase the versatility and robustness of forecasting systems. Stacking, boosting, and bagging techniques have successfully been applied within several frameworks, and have recently started to attract the interest of power system practitioners. The subject is therefore worthy of further investigation.

This book addresses the development of new, advanced, ensemble forecasting methods applied to power systems, collecting recent contributions to the development of accurate forecasts of energy-related variables by some of the most qualified experts in energy forecasting. Typical areas of research (renewable energy forecasting, load forecasting, energy price forecasting) are investigated, with relevant applications to the use of forecasts in energy management systems.

This book is relevant for forecasters and energy practitioners involved in power system planning and management. It also represents a good starting point for young researchers who want to discover potentialities of ensemble forecasting systems. Eventually, the contents of the book may be used by system operators and market operators in order to boost the performance of their in-house forecasting systems through ensemble refinement.

Antonio Bracale, Pasquale De Falco
Special Issue Editors

Article

Developing and Comparing Different Strategies for Combining Probabilistic Photovoltaic Power Forecasts in an Ensemble Method

Antonio Bracale [1], Guido Carpinelli [2,*] and Pasquale De Falco [1]

1. Department of Engineering, University of Napoli Parthenope, 80143 Naples, Italy; antonio.bracale@uniparthenope.it (A.B.); pasquale.defalco@uniparthenope.it (P.D.F.)
2. Department of Electrical Engineering and Information Technologies, University of Napoli Federico II, 80125 Naples, Italy
* Correspondence: guido.carpinelli@unina.it; Tel.: +39-081-768-3211

Received: 10 February 2019; Accepted: 11 March 2019; Published: 15 March 2019

Abstract: Accurate probabilistic forecasts of renewable generation are drivers for operational and management excellence in modern power systems and for the sustainable integration of green energy. The combination of forecasts provided by different individual models may allow increasing the accuracy of predictions; however, in contrast to point forecast combination, for which the simple weighted averaging is often a plausible solution, combining probabilistic forecasts is a much more challenging task. This paper aims at developing a new ensemble method for photovoltaic (PV) power forecasting, which combines the outcomes of three underlying probabilistic models (quantile k-nearest neighbors, quantile regression forests, and quantile regression) through a weighted quantile combination. Due to the challenges in combining probabilistic forecasts, the paper presents different combination strategies; the competing strategies are based on unconstrained, constrained, and regularized optimization problems for estimating the weights. The competing strategies are compared to individual forecasts and to several benchmarks, using the data published during the Global Energy Forecasting Competition 2014. Numerical experiments were run in MATLAB and R environments; the results suggest that unconstrained and Least Absolute Shrinkage and Selection Operator (LASSO)-regularized strategies exhibit the best performances for the datasets under study, outperforming the best competitors by 2.5 to 9% of the Pinball Score.

Keywords: forecast combination; photovoltaic power; probabilistic forecasting

1. Introduction

Load demand and non-controllable generation powers are the main sources of uncertainty in modern electrical grids and forecasting of these issues is of the greatest interest during planning and operation stages. In particular, disposing of accurate load and generation predictions is mandatory in order to tackle and solve a large variety of power system tasks, such as market bidding, energy dispatch in smart grids and microgrids, replacement reserve scheduling, virtual power plant aggregation, and sizing of battery energy storage systems [1–7].

Relevant literature on load and generation forecasting is quite heterogeneous; this is highlighted by the comparative dissertations in reviews and surveys [8,9], clearly showing that no method outperforms the others in every aspect. Major efforts have been devoted to point prediction, for which researchers and practitioners often individuate Artificial Neural Networks (ANN) [10,11], K-Nearest Neighbors (KNN) [12], support vector regression [13], Random Forests (RF) [14], and multiple linear regression models [15] as the best solutions.

Historically, research efforts have often been dedicated to load forecasting, since the generation has been mainly constituted by dispatchable fossil-fueled and hydroelectric plants. Today, instead, the widespread penetration of renewable generation, and in particular of photovoltaic (PV) and wind systems, makes forecasting renewable generation essential to cope with the new power system tasks, addressing the uncertainty of the energy source. Moreover, due to the intrinsic randomness of the physical phenomenon, probabilistic PV power forecasting is more adequate to deal with the management and operation of electrical networks under uncertainties [16,17]; however, only a minor part of the existing literature has dealt with PV power forecasting under a probabilistic framework. Relevant existing approaches are based on Quantile K-Nearest Neighbors (QKNN) [18], Quantile Regression Forests (QRFs) [19], Quantile Regression (QR) [20,21], and Gradient Boosting Regression Trees [18]; these models proved their effectiveness also in recent energy forecasting competitions [16,22], since the forecasting systems developed by the highest-ranking teams were based on these nonparametric probabilistic models.

New trends in probabilistic PV power forecasting individuate the probabilistic combination of individual forecasts as a suitable solution, in order to improve the accuracy of the results [9,20,23]. Probabilistic forecast combination is not as straightforward as it seems to be at first inspection. In contrast to combining point forecasts, for which the simple weighted averaging is often a plausible solution, combining probabilistic forecasts is a much more challenging task; the combined probabilistic forecasts indeed must retain adequate properties in terms of reliability and sharpness [20,24], and the main features of a probabilistic forecast (e.g., the ascending order of predictive quantiles) must be retained also by the combined forecasts [23].

Relevant literature has addressed these aspects under different points of view [9]. Individual probabilistic forecasts can be indeed merged: (i) by a combination of the predictive cumulative distribution functions [20]; or (ii) by a combination of the predictive quantiles [23]. The first combination type has already been applied to PV power forecasting, whereas the second combination type has yet to be applied to PV power forecasting (it has been presented in [23] only for load forecasting). Nevertheless, within these two types of approach, several strategies and architectures can be developed to combine forecasts, so there is room for further investigation and improvement.

In this context, this paper aims to provide a further contribution on the probabilistic combination of PV power forecasts. The paper develops and compares different forecast combination strategies applied to three individual probabilistic models (QKNN [18], QRF [19], and QR [20,21]), which are selected among the state-of-the-art nonparametric solutions for probabilistic forecasting. The outcomes of these models are properly combined under a competitive ensemble framework, based on a weighted combination of predictive quantiles.

Estimating the combination weights is a challenging task; several estimation strategies and architectures are therefore analyzed in this paper, in order to check the effectiveness of the model combination from different perspectives, and to allow picking the best solution for combining forecasts. In particular, the competing strategies are based on unconstrained, constrained, and regularized optimization problems for estimating the weights used to combine the predictive quantiles.

To guarantee the reproducibility of the experiments, the data published in the framework of the Global Energy Forecasting Competition 2014 (GEFCOM2014) [16] are used in this paper. In order to validate the proposal, the results are compared to relevant probabilistic benchmarks in actual scenarios.

Eventually, the main contributions of this paper are:

- the development of a new competitive ensemble method that combines the outcomes of three probabilistic models, selected among the ones which have proved consistency in probabilistic PV power forecasting;
- a critical analysis of different strategies and architectures to estimate the weights of the predictive quantile combination;
- a comparison of the results of the numerical experiments, based on the data published in the framework of the GEFCOM2014, with state-of-the-art probabilistic benchmarks.

The paper is organized as it follows. Section 2 provides an overview of the competitive ensemble method; Section 3 briefly describes the underlying probabilistic models; Section 4 presents the combination architectures and strategies analyzed in this paper; Section 5 shows benchmarks used for the comparison; Section 6 shows the results of the numerical experiments; and our conclusions are in Section 7.

2. Overview of the Proposed Competitive Ensemble Method for Forecasting PV Power

The forecasts of the underlying probabilistic models are combined in this paper in a competitive ensemble method, illustrated in Figure 1. Historical PV power and weather data, together with calendar qualitative variables, are the inputs of the procedure. These inputs are used by the underlying probabilistic models in order to build individual probabilistic forecasts of PV power, provided in terms of predictive quantiles. Eventually, the predictive quantiles returned from the underlying models are fed as inputs of the ensemble model, in order to be properly combined. In the forecast combination step, calendar variables may or may not be used; this differentiates the parameter estimation, as will be discussed in Section 4. The outputs of the procedure are probabilistic PV power forecasts, given in terms of predictive quantiles.

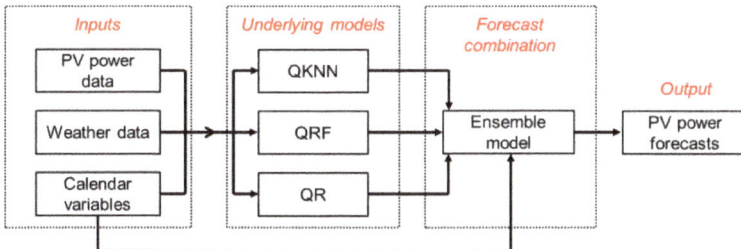

Figure 1. Overview of the proposed competitive ensemble method for forecasting photovoltaic (PV) power.

3. Probabilistic Underlying Models

Three probabilistic underlying models are selected and used to build individual forecasts. They are based on the QKNN, QRF, and QR techniques; all of these underlying models provide probabilistic forecasts in terms of predictive quantiles.

A brief description of these models is provided hereinafter. For each model, we assume that the same, following training data are available at the forecast origin t: (i) N historical values $P_{t-(N-1)}, P_{t-(N-2)}, \ldots, P_t$ of PV power; (ii) N vectors $z_{t-(N-1)}, z_{t-(N-2)}, \ldots, z_t$ of M predictors, corresponding to each of the N historical values of PV power. In particular, the generic jth vector of predictors is $z_j = \{z_{1_j}, \ldots, z_{M_j}\}$, for $j = t-(N-1), t-(N-2), \ldots, t$.

3.1. Quantile K-Nearest Neighbors

K-Nearest Neighbors (KNN) models are widely used in regression problems, due to their versatility and ease of use. Extending KNN to the probabilistic framework, thus formulating the QKNN model, is quite straightforward.

QKNN models are based on similarity effects. Given the predictors z_{t+k} (related to the forecast time horizon $t+k$, but known at the forecast origin t), the KNN model individuates, among the N predictor vectors, a subset $\mathcal{Z}_{t+k} = \{z_1^*, \ldots, z_K^*\}$ made of the K predictor vectors that are closest to the predictors z_{t+k}. In this paper, the proximity relationship is mathematically expressed using the Euclidean metric, defined as:

$$d(z_{t+k}, z_i) = \sqrt{\sum_{m=1}^{M} (z_{m_{t+k}}, z_{m_i})^2}. \tag{1}$$

The subset $\mathcal{P}_{t+k} = \{p_1^*, \ldots, p_K^*\}$ of measured PV powers, corresponding to the subset $\mathcal{Z}_{t+k}$, is straightforwardly individuated. The QKNN forecast $\hat{\mathcal{P}}_{t+k}^{(QKNN)}(q_i)$ at level q_i is then obtained as the sample q_i-quantile of the subset $\mathcal{P}_{t+k}$.

The hyper-parameter K (i.e., the number of neighbors) is selected in this paper in a cross-validation procedure.

3.2. Quantile Regression Forests

QRFs are groups of D decision trees, where individual trees are built by randomly selecting bagged subsets from the available pool of predictor variables.

Given the predictors z_{t+k} (related to the forecast time horizon $t+k$, but known at the forecast origin t), one leaf of each tree is univocally individuated. In particular, for the generic dth tree, it is denoted as $\mathcal{L}_d(z_{t+k})$. In QRF, all of the outcomes contained in the D leaves, which have been individuated, concur to form the probabilistic forecast for the time horizon $t+k$.

The QRF predictive distribution is estimated as:

$$\hat{F}(P_{t+k} \leq P^* | z_{t+k}) = \sum_{n=1}^{N} w_n(z_{t+k}) \mathbf{1}\left\{P_{t-(n-1)} \leq P^*\right\}, \quad (2)$$

where $\mathbf{1}\{\cdot\} = 1$ if the condition in the brackets is true, $\mathbf{1}\{\cdot\} = 0$ if the condition is false, and a weight coefficient $w_n(z_{t+k})$ is estimated for each of the N historical vectors of predictors, as:

$$w_n(z_{t+k}) = \frac{1}{D} \sum_{d=1}^{D} \frac{\mathbf{1}\left\{z_{t-(n-1)} \in R_{\mathcal{L}_d(z_{t+k})}\right\}}{\sum_{n=1}^{N} \mathbf{1}\left\{z_{t-(n-1)} \in R_{\mathcal{L}_d(z_{t+k})}\right\}}, \quad (3)$$

and $R_{\mathcal{L}_d(z_{t+k})}$ is the rectangular subspace of $\mathbb{R}^M$ in which the leaf $\mathcal{L}_d(z_{t+k})$ finds its values.

Obtaining the QRF forecast $\mathcal{P}_{t+k}^{(QRF)}(q_i)$ at level q_i is straightforward from (2); it is:

$$\hat{\mathcal{P}}_{t+k}^{(QRF)}(q_i) = \inf\{P^* : \hat{F}(P_{t+k} \leq P^* | z_{t+k}) \geq q_i\}. \quad (4)$$

The hyper-parameter D (i.e., the number of trees in the forest) is selected in this paper in a cross-validation procedure.

3.3. Quantile Regression

QR is a multiple linear regression model, the parameters of which are not estimated in a traditional ordinary least square approach, but instead they are estimated by minimizing the Pinball Score (PS) [16,25] in the training period. The PS is a proper score [25], which simultaneously accounts for reliability and sharpness of the probabilistic forecasts; it is the most common index in evaluating probabilistic forecasts, and is therefore used in all of the comparative analyses in this paper.

The analytic formulation of the QR model is:

$$P_{t+k} = z_{t+k} \beta^{(q_i)} + \varepsilon_{t+k}^{(q_i)}, \quad (5)$$

where $\beta^{(q_i)}$ is the vector of parameters to be estimated, and $\varepsilon_{t+k}^{(q_i)}$ is the residual. Parameters are estimated from:

$$\hat{\beta}^{(q_i)} = \underset{\beta^{(q_i)}}{\operatorname{argmin}} \sum_{n=1}^{N} \varepsilon_{t-(n-1)}^{(q_i)} \cdot (q_i - \mathbf{1}\{\varepsilon_{t-(n-1)}^{(q_i)} < 0\}). \quad (6)$$

The unconstrained nonlinear programming problem in (6) can be put in a constrained linear programming problem [26]; this allows to increase the computational efficiency. It is represented as:

$$\hat{\beta}^{(q_i)}, \hat{\varepsilon}^{(q_i)^+}, \hat{\varepsilon}^{(q_i)^-} = \underset{\beta^{(q_i)}, \varepsilon^{(q_i)^+}, \varepsilon^{(q_i)^-}}{\operatorname{argmin}} q_i \mathbf{1}_{[Nx1]} \varepsilon^{(q_i)^+} + (1-q_i) \mathbf{1}_{[Nx1]} \varepsilon^{(q_i)^-},$$
$$\text{s.t. } z_{t-(n-1)} \beta^{(q_i)} + \varepsilon^{(q_i)^+}_{t-(n-1)} - \varepsilon^{(q_i)^-}_{t-(n-1)}, \ \forall n, \qquad (7)$$
$$\varepsilon^{(q_i)^+}_{t-(n-1)}, \varepsilon^{(q_i)^-}_{t-(n-1)} \geq 0, \ \forall n,$$

where $\mathbf{1}_{[Nx1]}$ is a $[Nx1]$ vector of ones, and:

$$\varepsilon^{(q_i)^+}_{t-(n-1)} = \varepsilon^{(q_i)}_{t+k} \cdot \mathbf{1}\left\{\varepsilon^{(q_i)}_{t-(n-1)} \geq 0\right\}, \qquad (8)$$

$$\varepsilon^{(q_i)^-}_{t-(n-1)} = -\varepsilon^{(q_i)}_{t+k} \cdot \mathbf{1}\left\{\varepsilon^{(q_i)}_{t-(n-1)} < 0\right\}. \qquad (9)$$

The QR forecast $\hat{P}^{(QR)}_{t+k}(q_i)$ at level q_i is then obtained as:

$$\hat{P}^{(QR)}_{t+k}(q_i) = z_{t+k}\hat{\beta}^{(q_i)}. \qquad (10)$$

4. The Competitive Ensemble Model for Forecast Combination

The competitive ensemble model is based on the Quantile Weighted Sum (QWS), which has recently been applied to probabilistic load forecasting with interesting results [23]. Eight different strategies are proposed and compared in this paper:

- the Pure Quantile Weighted Sum (PQWS);
- the Hourly Quantile Weighted Sum (HQWS);
- the Pure Constrained Quantile Weighted Sum (PCQWS);
- the Hourly Constrained Quantile Weighted Sum (HCQWS);
- the Pure Quantile Weighted Sum with Least Absolute Shrinkage and Selection Operator (LASSO) Regularization (PQWSLR);
- the Hourly Quantile Weighted Sum with LASSO Regularization (HQWSLR);
- the Pure Quantile Weighted Sum with Ridge Regularization (PQWSRR);
- the Hourly Quantile Weighted Sum with Ridge Regularization (HQWSRR).

In the "pure" approaches, weights are estimated without any differentiation in terms of daily periodicity, whereas in the "hourly" approaches weights are estimated using only same-hour observations; thus the weights are differentiated by the hour of the day to account for the daily periodicity of the PV power pattern. The last four approaches are extended in this paper starting from the Least Absolute Shrinkage and Selection Operator (LASSO) quantile regression [27] and from the Ridge quantile regression [28], respectively, which allow regularizing the weights by assigning a penalty linked to the magnitude of the weights.

In the PQWS and HQWS strategies, the weights are estimated without any constraint or regularization loss. Compared to the constrained or regularized strategies, the PQWS and HQWS strategies return the smallest in-sample PS, since the minimization problem is unconstrained. However, there is no assurance that these weights are the best picks for out-of-sample forecasts. This is a common issue for regression applied to forecasting, in which overfitting the training data has negative consequences when the model is used to forecast unknown data.

Therefore, in this paper, we compare the results of the unconstrained, non-regularized strategies to constrained and regularized strategies, in order to check their performances and to pick the best strategy to be used in practical applications.

The PCQWS and the HCQWS are the constrained strategies, in which weights are forced to sum for the unity. This ensures that the predictive quantiles of the combined forecasts do not oddly deviate from the average value of the three predictive quantiles of the individual predictors. The in-sample PS of the PCQWS (HCQWS) strategy is obviously greater than the corresponding in-sample PS of the PQWS (HQWS) strategy, but the out-of-sample performances may be very different.

The PQWSLR, the HQWSLR, the PQWSRR, and the HQWSRR strategies were instead developed in order to estimate weights having a regularized magnitude (in absolute value). Indeed, regularization of the parameters is a well-known strategy in order to avoid overfitting by penalizing the returned objective function (in this case, the PS), adding a loss term which directly depends on the magnitude of the parameters. In this paper, both the LASSO and the Ridge regularization are tested, in order to provide a comprehensive analysis. Note that, in these cases, the in-sample PSs of the PQWSLR/PQWSRR (HQWSLR/HQWSRR) strategies are obviously greater than the corresponding in-sample PS of the PQWS (HQWS) strategy, but the out-of-sample performances may be very different.

All of the strategies developed in this paper are presented in the following subsections.

4.1. Pure Quantile Weighted Sum

PQWS combination returns predictive quantiles at a given level q_i, by summing the predictive quantiles (at the same level q_i) of the underlying models, multiplied for coefficients $\omega^{(q_i)} = \left\{\omega_1^{(q_i)}, \omega_2^{(q_i)}, \omega_3^{(q_i)}\right\}$ that are estimated in the training step.

Starting from the PQWS approach, two strategies are separately analyzed in this paper: a combination of all of the three individual forecasts (PQWS3) and a combination of the two best individual forecasts (PQWS2). This differentiated analysis is run in order to check whether the addition of a third individual forecast, which is clearly worse than the other two, may add useful information when building the ensemble. In the following formulation, we will refer to the PQWS3 strategy, since its extension to the PQWS2 is trivial. The model is:

$$P_{t+k}^{(PQWS3)}(q_i) = \omega_1^{(q_i)} \hat{P}_{t+k}^{(QKNN)}(q_i) + \omega_2^{(q_i)} \hat{P}_{t+k}^{(QRF)}(q_i) + \omega_3^{(q_i)} \hat{P}_{t+k}^{(QR)}(q_i). \tag{11}$$

The weights are estimated from:

$$\hat{\omega}^{(q_i)} = \left\{\hat{\omega}_1^{(q_i)}, \hat{\omega}_2^{(q_i)}, \hat{\omega}_3^{(q_i)}\right\} = \underset{\omega^{(q_i)}}{\mathrm{argmin}} \sum_{l=1}^{L} \left[P_l - P_l^{(PQWS3)}(q_i)\right] \cdot (q_i - 1\{P_l < P_l^{(PQWS3)}(q_i)\}) \tag{12}$$

that is, by minimizing the PS in the training interval, which is made of L observed points.

The hyper-parameter L is the length of the dataset used to train the weights of the combination models; it could be optimized by means of a model selection procedure (e.g., cross-validation). However, no selection procedure was run in this paper to pick the optimal hyper-parameter L; our purpose was instead to provide an exhaustive comparative analysis on the variation of the forecast errors with respect to this hyper-parameter. Nevertheless, the results of the comparative analysis can be used by the forecaster to build subsequent out-of-sample forecasts, picking the optimal result.

4.2. Hourly Quantile Weighted Sum

The daily seasonality of the PV power time series is taken into account in the HQWS approach. For the same purposes enunciated beforehand, two strategies were developed from the HQWS approach and separately analyzed: a combination of all of the three individual forecasts (HQWS3) and a combination of the two best individual forecasts (HQWS2). We present the HQWS3 strategy, since the extension to the HQWS2 case is trivial. The model is:

$$P_{t+k}^{(PQWS3)}(q_i) = \sum_{h=1}^{24}\left[\omega_{h_1}^{(q_i)} \hat{P}_{t+k}^{(QKNN)}(q_i) + \omega_{h_2}^{(q_i)} \hat{P}_{t+k}^{(QRF)}(q_i) + \omega_{h_3}^{(q_i)} \hat{P}_{t+k}^{(QR)}(q_i)\right] \cdot hod_{t+k}^{(h)}, \tag{13}$$

where $hod_{t+k}^{(h)} = 1$ if the forecast horizon $t+k$ is the hth hour of the day, and $hod_{t+k}^{(h)} = 0$ otherwise. The weights are estimated from:

$$\hat{\omega}^{(q_i)} = \left\{\hat{\omega}_{1_1}^{(q_i)}, \hat{\omega}_{1_2}^{(q_i)}, \hat{\omega}_{1_3}^{(q_i)}, \ldots, \hat{\omega}_{24_1}^{(q_i)}, \hat{\omega}_{24_2}^{(q_i)}, \hat{\omega}_{24_3}^{(q_i)}\right\} =$$
$$= \underset{\omega^{(q_i)}}{\operatorname{argmin}} \sum_{l=1}^{L} \left[P_l - P_l^{(PQWS3)}(q_i)\right] \cdot (q_i - 1\{P_l < P_l^{(PQWS3)}(q_i)\}). \quad (14)$$

The daily periodicity of the PV power pattern is therefore also accounted for in the forecast combination; Equation (14) is a new formulation proposed in this paper to account for it in PV power forecast combination. This new HQWS approach is indeed expected to improve the forecast combination, by differentiating the weights not only for different quantiles but also for different hours of the day.

4.3. Pure and Hourly Constrained Quantile Weighted Sum

The PCQWS and the HCQWS approaches are based on a constrained optimization formulation, in which the sum of the weights is constrained to the unity. Also, for these approaches we differentiate between a combination of all of the three individual forecasts (PCQWS3 and HCQWS3 strategies) and a combination of the two best individual forecasts (PCQWS2 and HCQWS2 strategies).

For the PCQWS3 strategy, the model is analogous to Equation (11), but the weights are estimated from:

$$\hat{\omega}^{(q_i)} = \left\{\hat{\omega}_1^{(q_i)}, \hat{\omega}_2^{(q_i)}, \hat{\omega}_3^{(q_i)}\right\} = \underset{\omega^{(q_i)}}{\operatorname{argmin}} \sum_{l=1}^{L} \left[P_l - P_l^{(PQWS3)}(q_i)\right] \cdot (q_i - 1\{P_l < P_l^{(PQWS3)}(q_i)\})$$
$$\text{s.t. } \omega_1^{(q_i)} + \omega_2^{(q_i)} + \omega_3^{(q_i)} = 1. \quad (15)$$

For the HCQWS3 strategy, the model is analogous to Equation (13), but the weights are estimated from:

$$\hat{\omega}^{(q_i)} = \left\{\hat{\omega}_{1_1}^{(q_i)}, \hat{\omega}_{1_2}^{(q_i)}, \hat{\omega}_{1_3}^{(q_i)}, \ldots, \hat{\omega}_{24_1}^{(q_i)}, \hat{\omega}_{24_2}^{(q_i)}, \hat{\omega}_{24_3}^{(q_i)}\right\} =$$
$$= \underset{\omega^{(q_i)}}{\operatorname{argmin}} \sum_{l=1}^{L} \left[P_l - P_l^{(PQWS3)}(q_i)\right] \cdot (q_i - 1\{P_l < P_l^{(PQWS3)}(q_i)\})$$
$$\text{s.t. } \omega_{1_1}^{(q_i)} + \omega_{1_2}^{(q_i)} + \omega_{1_3}^{(q_i)} = 1 \quad (16)$$
$$\vdots$$
$$\omega_{24_1}^{(q_i)} + \omega_{24_2}^{(q_i)} + \omega_{24_3}^{(q_i)} = 1.$$

4.4. Quantile Weighted Sum with LASSO Regularization

The PQWSLR and the HQWSLR approaches are based on the regularization of the weights through the LASSO [27]. In contrast to the constrained approaches, in which the sum of the weights is assigned, the regularization of parameters in the PQWSLR and HQWSLR approaches is an output of the model itself (which indeed requires no pre-assignment from the forecaster). Due to the intrinsic capability of the LASSO in reducing the impact of uninformative predictors by assigning smaller (or even zero) weights to them [16], these two approaches were developed and tested only for the combination of all three of the individual forecasts, thus each of them straightforwardly identifies one combination strategy.

For the PQWSLR strategy, the model is analogous to Equation (11), but the weights are estimated from:

$$\hat{\omega}^{(q_i)} = \left\{\hat{\omega}_1^{(q_i)}, \hat{\omega}_2^{(q_i)}, \hat{\omega}_3^{(q_i)}\right\} =$$
$$= \underset{\omega^{(q_i)}}{\operatorname{argmin}} \sum_{l=1}^{L} \left[P_l - P_l^{(PQWS3)}(q_i)\right] \cdot (q_i - 1\{P_l < P_l^{(PQWS3)}(q_i)\}) + \lambda_L \sum_{j=1}^{3} |\omega_j|. \quad (17)$$

For the HQWSLR strategy, the model is analogous to Equation (13), but the weights are estimated from:

$$\hat{\omega}^{(q_i)} = \{\hat{\omega}_{1_1}^{(q_i)}, \hat{\omega}_{1_2}^{(q_i)}, \hat{\omega}_{1_3}^{(q_i)}, \ldots, \hat{\omega}_{24_1}^{(q_i)}, \hat{\omega}_{24_2}^{(q_i)}, \hat{\omega}_{24_3}^{(q_i)}\} =$$
$$= \underset{\omega^{(q_i)}}{\arg\min} \sum_{l=1}^{L} \left[P_l - P_l^{(PQWS3)}(q_i)\right] \cdot (q_i - 1\{P_l < P_l^{(PQWS3)}(q_i)\}) + \lambda_L \sum_{j=1}^{3} \sum_{u=1}^{24} \left|\omega_{u_j}\right| \quad (18)$$

The selection of the penalty coefficient λ_L (which is an important hyper-parameter in LASSO regression) is performed in this paper in 5-fold cross-validation.

4.5. Quantile Weighted Sum with RIDGE Regularization

The PQWSRR and HQWSRR approaches are based on the Ridge regularization [28]. The models are very similar to the LASSO-based ones, and also, in this case, the intrinsic capability in reducing the impact of uninformative predictors by assigning smaller weights to them [28] lead us to develop and test the PQWSRR and HQWSRR approaches only for the combination of all of the three individual forecasts, developing one strategy for each approach.

For the PQWSRR strategy, the model is analogous to Equation (11), but the weights are estimated from:

$$\hat{\omega}^{(q_i)} = \{\hat{\omega}_1^{(q_i)}, \hat{\omega}_2^{(q_i)}, \hat{\omega}_3^{(q_i)}\} =$$
$$= \underset{\omega^{(q_i)}}{\arg\min} \sum_{l=1}^{L} \left[P_l - P_l^{(PQWS3)}(q_i)\right] \cdot (q_i - 1\{P_l < P_l^{(PQWS3)}(q_i)\}) + \lambda_R \sum_{j=1}^{3} \omega_j^2. \quad (19)$$

For the HQWSRR strategy, the model is analogous to Equation (13), but the weights are estimated from:

$$\hat{\omega}^{(q_i)} = \{\hat{\omega}_{1_1}^{(q_i)}, \hat{\omega}_{1_2}^{(q_i)}, \hat{\omega}_{1_3}^{(q_i)}, \ldots, \hat{\omega}_{24_1}^{(q_i)}, \hat{\omega}_{24_2}^{(q_i)}, \hat{\omega}_{24_3}^{(q_i)}\} =$$
$$= \underset{\omega^{(q_i)}}{\arg\min} \sum_{l=1}^{L} \left[P_l - P_l^{(PQWS3)}(q_i)\right] \cdot (q_i - 1\{P_l < P_l^{(PQWS3)}(q_i)\}) + \lambda_R \sum_{j=1}^{3} \sum_{u=1}^{24} \omega_{u_j}^2. \quad (20)$$

The selection of the penalty coefficient λ_R (which is an important hyper-parameter in Ridge regression) is performed in this paper in 5-fold cross-validation.

5. Benchmarks

The ensemble combination of probabilistic individual forecasts was mainly assessed in terms of relative improvement with respect to individual predictions. However, three relevant benchmarks were also added for comparison. They are briefly recalled in the following subsections.

5.1. Naïve Benchmark

A Naïve Benchmark (NB), was provided by the GEFCOM2014 organizers [16]. It consists of point forecasts which are repeated for the 99 predictive quantiles. This benchmark was added in this paper in order to provide a direct comparison with outcomes of the GEFCOM2014.

5.2. Quantile Artificial Neural Network

An ANN-based probabilistic benchmark (QANN) is the second benchmark, which was added to provide a comparison with an artificial intelligence technique. The QANN consists of a feedforward neural network, which was trained upon the 70% of the available training data by minimizing the PS using a particle-swarm optimization algorithm. The hyper-parameter optimization was performed on the remaining 30% of the available training data, reserved for validation. A dedicated neural network was trained for each predictive quantile level, in order to improve the performances. The QANN was performed by the neural network toolbox in MATLAB.

5.3. Gradient Boosting Regression Trees

A Gradient-Boosting Regression Tree (GBRT) benchmark, which was added due to the great performances showed in the winner methods during the GEFCOM2014. Also, in this case a dedicated model was trained for each quantile level. The GBRT was developed using the gbm package in R [29].

5.4. Bayesian Method

A Bayesian (BAY) benchmark was adapted from the methods presented in [20,30], in order to suit the forecasting scheme of the GEFCOM2014. In particular, the underlying deterministic model selected to forecast the expected values of the posterior predictive distributions consists of an average of a GBRT model and of a RF model; exogenous time series approaches indeed performed quite poorly, due to the monthly forecast horizons. The BAY benchmark is an hybrid parametric model, and it was specifically added in the comparative analysis in order to also provide a parametric reference for the results.

6. Numerical Application

The strategies for combining individual probabilistic forecasts are quantitatively assessed in this Section, using actual PV power data provided in the context of an energy forecasting competition [16]. First, we present the data used for the numerical experiments and the accuracy of the results of individual forecasts; later we assess the accuracy of the forecast combination strategies. The PS values are used to quantitatively estimate the forecast performances [16,23].

6.1. Characteristics of the Data

The PV power data refers to three zones, which are geographically correlated; each time series was collected in a time interval ranging from April 2012 to June 2014. For reproducibility, we follow the same division kept by the organizers of the competition: the first year of data (April 2012–March 2013) was used only for training the underlying models; each of the remaining 15 months (April 2013–June 2014) constitutes a forecasting task. In order to improve the performances of the underlying models and of the forecast combination, and to maintain consistency between the outcomes of different forecast approaches, we selected a 1-year constant-length window for training underlying models at each task; the window shifts towards the most recent task.

The forecast combination is trained upon different numbers of tasks (i.e., using a different hyper-parameter L). Also in this case, once the hyper-parameter L is iteratively assigned, the time window used for training the combination weights has a constant length, and it shifts towards the most recent task. We reserve the last 5 tasks (February 2014–June 2014) to test the out-of-sample performances.

Table 1 shows the main statistical properties (mean, median, and variance) of the three PV datasets considered, as a whole. Note that all of the data provided by the competition organized are normalized. More details can be found in [16].

Table 1. Statistical properties of the PV data considered.

Zone	Statistical Parameter [-]		
	Mean	Median	Standard Deviation
1	0.1693	0.0026	0.2588
2	0.1879	0.0022	0.2756
3	0.1939	0.0028	0.2821

6.2. Assessment of the Accuracy of Individual Forecasts

We investigated the accuracy of the individual forecasts in all of the 15 tasks. The results during the last 5 tasks were also used as benchmarks, to compare the performances of the forecast combination approaches in the test step. Figures 2–4 show the plots of the PSs obtained using the QKNN, the QRF, and the QR, to the 15 considered tasks, for the zones 1, 2, and 3, respectively. The benchmark PS values of the QANN, of the GBRT, of the BAY, and of the NB are also shown as a reference.

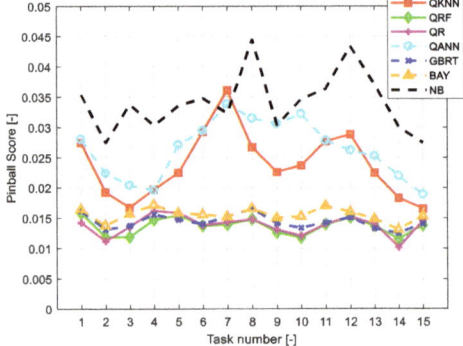

Figure 2. Pinball Score values of individual forecasts for zone 1.

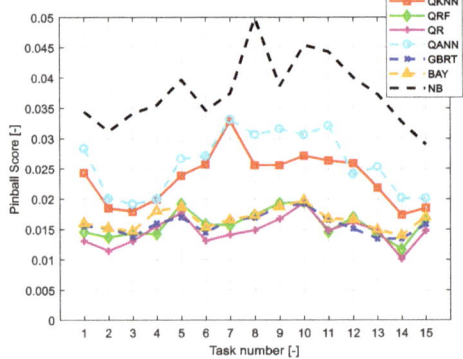

Figure 3. Pinball Score values of individual forecasts for zone 2.

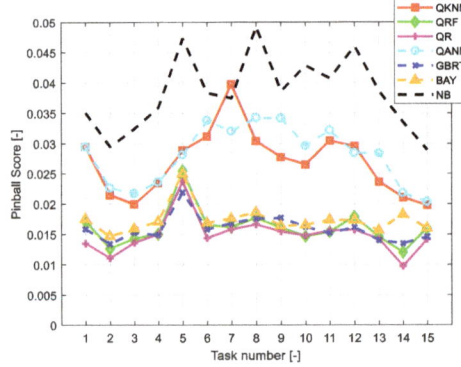

Figure 4. Pinball Score values of individual forecasts for zone 3.

Figures 2–4 clearly highlight the superior performances of QRF, QR, GBRT, and BAY models, with respect to the other models. For zone 1, QRF and QR perform very similarly, whereas for zones 2 and 3 the QR on average outperforms the QRF. The GBRT benchmark exhibits performances, on average, slightly worse than the QRF and QR; however, it outperforms the QANN and the QKNN for all of the zones considered. The BAY benchmark is, on average, slightly worse than the QRF, QR, and GBRT, whereas it outperforms QKNN, QANN, and NB.

Table 2 shows the PS values of the individual forecasts and of the NB forecasts, averaged over tasks 11–15 (i.e., the tasks reserved for comparing the out-of-sample combination results). The PS values in Table 2 confirm the considerations made on the basis of the graphical inspection of Figures 2–4.

Table 2. Pinball Score values averaged over tasks 11–15.

Method	Pinball Score [-]		
	Zone 1	Zone 2	Zone 3
QKNN	0.0228	0.0220	0.0249
QRF	0.0136	0.0148	0.0152
QR	0.0136	0.0141	0.0139
QANN	0.0240	0.0243	0.0262
GBRT	0.0138	0.0149	0.0147
BAY	0.0153	0.0159	0.0169
NB	0.0349	0.0367	0.0376

6.3. Assessment of the Accuracy of Combined Forecasts

PQWS2, PQWS3, HQWS2, HQWS3, PCQWS2, PCQWS3, HCQWS2, HCQWS3, PQWSLR, HQWSLR, PQWSRR, and HQWSRR forecasts are analyzed in this sub-Section. Different values of hyper-parameter L (i.e., the length of the dataset used to train the weights of the combination models) are considered separately; in particular, they cover the 1, 2, ... , 10 most recent tasks.

Figures 5–7 show, for zones 1, 2, and 3, the PS values of the forecasts to the number of tasks considered to form the individual forecast dataset. In particular, Figures 5, 6 and 7a illustrate the PQWS2, PQWS3, HQWS2, HQWS3 results; Figures 5, 6 and 7b illustrate the PCQWS2, PCQWS3, HCQWS2, HCQWS3 results; and Figures 5, 6 and 7c illustrate the PQWSLR, HQWSLR, PQWSRR, HQWSRR results.

These figures clearly highlight that the PQWS3 outperforms the unconstrained, non-regularized strategies for all of the tasks considered. For zone 1, the HCQWS3 outperforms the other constrained strategies, whereas PCQWS3 performs better than the other constrained strategy for zone 3. Note, however, that the constrained strategies, compared to the unconstrained and regularized strategies, have quite similar results for zones 1 and 2, whereas the constrained strategies are definitely less accurate than the unconstrained and the regularized strategies for zone 3.

The smallest error score among all of the options considered is obtained for the zones 1 and 2 by using the HQWSLR with 7 tasks in the individual forecast dataset, whereas the best forecasts among all of the options considered for zone 3 are obtained through the PQWS3 with 6 tasks in the individual forecast dataset.

The trends of the PS of the combined forecasts are quite similar, as the performances significantly increase by using more than four tasks to form the individual forecast dataset; this improvement is at maximum around 6–8 tasks, and it slightly decreases with more tasks.

Altogether, the PQWS approaches outperform the HQWS ones, thus a more general model works better than a model with too much differentiation in the unconstrained, non-regularized estimation. Things change when the weight estimation is subject to constraints or to regularization; the hourly differentiation improves the performances of the forecasting ensemble for zones 1 and 2.

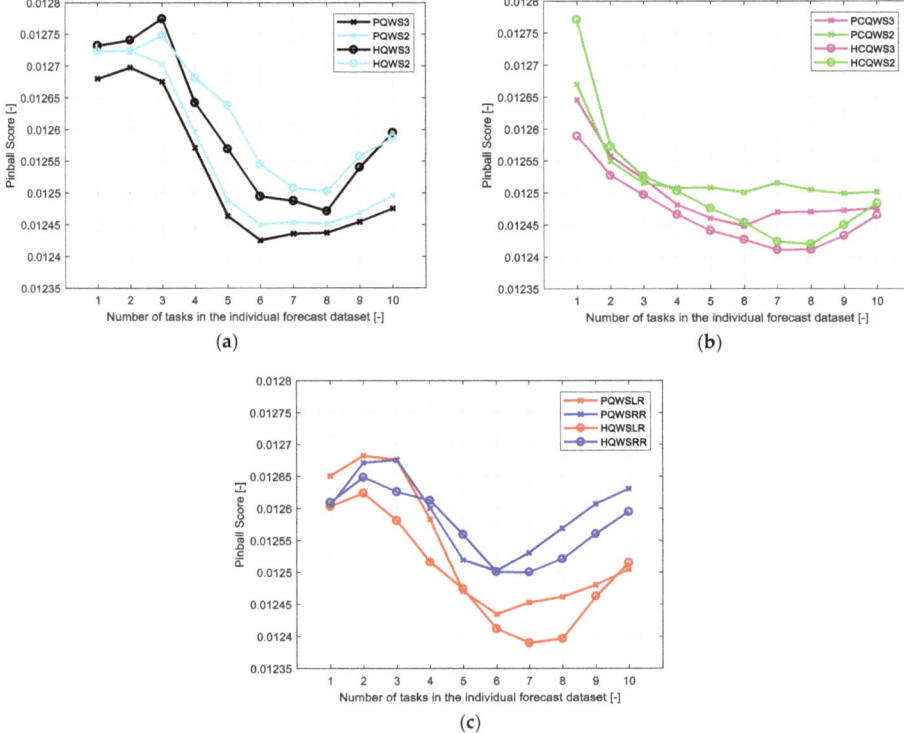

Figure 5. Pinball Score values of combined forecasts for the zone 1: (**a**) PQWS2, PQWS3, HQWS2, and HQWS3 strategies; (**b**) PCQWS2, PCQWS3, HCQWS2, HCQWS3 strategies; (**c**) PQWSLR, HQWSLR, PQWSRR, HQWSRR strategies.

We now analyze in detail the best competitors for the three zones: the HQWSLR for zone 1 and 2, and the PQWS3 for zone 3. In particular, for both strategies we consider only the best-case scenario and the worst-case scenario, in terms of number of tasks in the individual forecast dataset. For the PQWS3, the optimal number of tasks selected to train upon the individual forecast datasets was 6, 8, and 6, for zones 1, 2, and 3, respectively; the worst performances of the PQWS3 were instead obtained with 2, 1, and 2 tasks for zones 1, 2, and 3, respectively. For the HQWSLR, the optimal number of tasks selected to train upon the individual forecast datasets was 7 for all of the zones; the worst performances of the HQWSLR were instead obtained with 2, 1, and 1 tasks for zones 1, 2, and 3, respectively.

The comprehensive Table 3 compares the corresponding PS values of these best- and worst-case scenarios to the ones obtained for the individual forecasts and for the benchmark (see Table 2). It is evident from these results that the forecast combination, either through the PQWS3 or the HQWSLR, always improves the skill of the forecasts.

We quantitatively assessed the results of the PQWS3 and of the HQWSLR by comparing them to the most competitive benchmarks for each zone, which are the QRF for the zone 1, and the QR for the zones 2 and 3. In the best-case scenario, the PS obtained through the PQWS3 is about 9%, 3.5%, and 6.5% smaller than the corresponding PS of the most competitive benchmark for zones 1, 2, and 3, respectively; in the worst-case scenario, the PS obtained through the PQWS3 is about 6.5%, 2%, and 5% smaller the corresponding PS of the most competitive benchmark for zones 1, 2, and 3, respectively. The improvement of the HQWSLR towards the most competitive benchmark for each zone is instead about 9%, 4.5%, and 6.5% in the best-case scenario, and about 7.5%, 2.5%, and 4.5% in the worst-case scenario, for zones 1, 2, and 3 respectively.

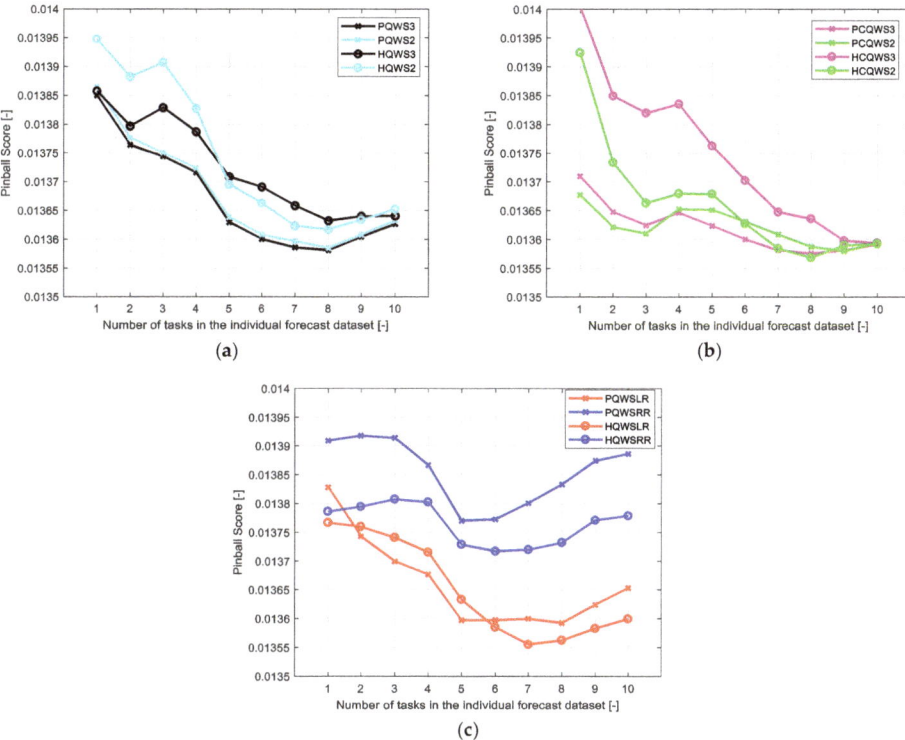

Figure 6. Pinball Score values of combined forecasts for the zone 2: (**a**) the Pure Quantile Weighted Sum (PQWS)2, PQWS3, Hourly Quantile Weighted Sum (HQWS)2, and HQWS3 strategies; (**b**) Pure Constrained Quantile Weighted Sum (PCQWS)2, PCQWS3, Hourly Constrained Quantile Weighted Sum (HCQWS)2, HCQWS3 strategies; (**c**) Pure Quantile Weighted Sum with Least Absolute Shrinkage and Selection Operator (LASSO) Regularization (PQWSLR), Hourly Quantile Weighted Sum with LASSO Regularization (HQWSLR), Pure Quantile Weighted Sum with Ridge Regularization (PQWSRR), Hourly Quantile Weighted Sum with Ridge Regularization (HQWSRR) strategies.

Table 3. Pinball Score values averaged over the tasks 11–15. Bold values highlight the best results for each zone.

Method	Pinball Score [-]		
	Zone 1	Zone 2	Zone 3
QKNN	0.0228	0.0220	0.0249
QRF	0.0136	0.0148	0.0152
QR	0.0136	0.0141	0.0139
QANN	0.0240	0.0243	0.0262
GBRT	0.0138	0.0149	0.0147
BAY	0.0153	0.0159	0.0169
NB	0.0349	0.0367	0.0376
Best PQWS3	0.0124	0.0136	**0.0130**
Worst PQWS3	0.0127	0.0138	0.0132
Best HQWSLR	**0.0124**	**0.0135**	0.0130
Worst HQWSLR	0.0126	0.0138	0.0133

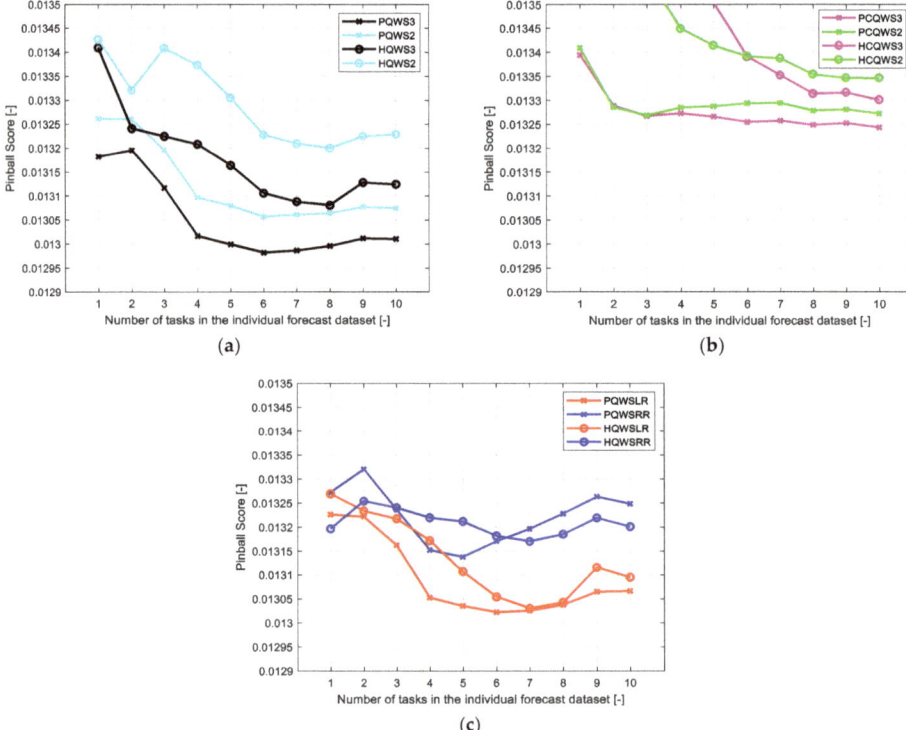

Figure 7. Pinball Score values of combined forecasts for the zone 3: (**a**) PQWS2, PQWS3, HQWS2, and HQWS3 approaches; (**b**) PCQWS2, PCQWS3, HCQWS2, HCQWS3 approaches; (**c**) PQWSLR, HQWSLR, PQWSRR, HQWSRR approaches.

7. Conclusions

This paper discusses several strategies that have been developed to combine individual probabilistic PV power forecasts, aimed at building combined forecasts which are more accurate than individual predictions. Several types of combination strategies and architectures were developed in a competitive ensemble framework; all of them are based on the weighted quantile combination. The proposal was validated through numerical experiments based on PV power data published during the GEFCOM2014; several benchmarks are also presented, in order to compare the results.

The comparison of different forecast strategies for three different generation zones suggests that:

- the weighted quantile combination was effective in improving the accuracy of forecasts; it is able to outperform the accuracy of individual probabilistic forecasts, which is the main aim of competitive ensemble methods.
- The forecast combination improved the skill of the forecasts in all of the scenarios considered, with a reduction in terms of PS that is up to 9%.
- On average, the best results were obtained using the HQWSLR combination strategy for zones 1 and 2, and the PQWS3 combination strategy for the zone 3; the optimal length of the dataset used to train the weights of the combination models always ranges between 6 and 8 tasks.
- Adding the forecasts of an individual model which has worse performances than the other individual models appears to provide useful diversity in the ensemble approach; this appears to be valid both for unconstrained, non-regularized strategies and for constrained strategies.

- Adding too much dispersion to the forecast combination by estimating weights for each hour of the day does not improve the quality of the results for unconstrained, non-regularized regression; vice versa, constraints and/or regularization allow taking benefit from this hourly differentiation.

Author Contributions: Conceptualization, A.B., G.C., and P.D.F.; methodology, A.B., G.C., and P.D.F.; software, A.B., G.C., and P.D.F.; validation, A.B., G.C., and P.D.F.; formal analysis, A.B., G.C., and P.D.F.; investigation, A.B., G.C., and P.D.F.; resources, A.B., G.C., and P.D.F.; data curation, A.B., G.C., and P.D.F.; writing—original draft preparation, A.B., G.C., and P.D.F.; writing—review and editing, A.B., G.C., and P.D.F.; visualization, A.B., G.C., and P.D.F.; supervision, A.B., G.C., and P.D.F.

Funding: This research received no external funding.

Acknowledgments: This paper was in part supported by the University of Napoli Parthenope in the framework of the research project "La Power Quality nelle Microgrids", and by the European Union's Horizon 2020 research and innovation programme under grant agreement n°773406 (OSMOSE - Optimal System-Mix Of flexibility Solutions for European electricity).

Conflicts of Interest: The authors declare no conflict of interest.

References

1. Toubeau, J.F.; Bottieau, J.; Vallee, F.; De Greve, Z. Deep learning-based multivariate probabilistic forecasting for short-term scheduling in power markets. *IEEE Trans. Power Syst.* **2018**, *34*, 1203–1215. [CrossRef]
2. Javadi, M.; Marzband, M.; Funsho Akorede, M.; Godina, R.; Saad Al-Sumaiti, A.; Pouresmaeil, E. A centralized smart decision-making hierarchical interactive architecture for multiple home microgrids in retail electricity market. *Energies* **2018**, *11*, 3144. [CrossRef]
3. Marzband, M.; Azarinejadian, F.; Savaghebi, M.; Pouresmaeil, E.; Guerrero, J.M.; Lightbody, G. Smart transactive energy framework in grid-connected multiple home microgrids under independent and coalition operations. *Renew. Energy* **2018**, *126*, 95–106. [CrossRef]
4. Capizzi, G.; Lo Sciuto, G.; Napoli, C.; Tramontana, E. Advanced and adaptive dispatch for smart grids by means of predictive models. *IEEE Trans. Smart Grid* **2018**, *9*, 6684–6691. [CrossRef]
5. Matos, M.; Bessa, R.J.; Botterud, A.; Zhou, Z. Forecasting and setting power system operating reserves. In *Renewable Energy Forecasting: From Models to Applications*, 1st ed.; Kariniotakis, G., Ed.; Woodhead Publishing: Duxford, UK, 2017; pp. 279–308.
6. Camal, S.; Michiorri, A.; Kariniotakis, G. Optimal offer of automatic frequency restoration reserve from a combined PV/wind virtual power plant. *IEEE Trans. Power Syst.* **2018**, *33*, 6155–6170. [CrossRef]
7. Carpinelli, G.; Mottola, F.; Proto, D. Probabilistic sizing of battery energy storage when time-of-use pricing is applied. *Electr. Power Syst. Res.* **2016**, *141*, 73–83. [CrossRef]
8. Van der Meer, D.W.; Widén, J.; Munkhammar, J. Review on probabilistic forecasting of photovoltaic power production and electricity consumption. *Renew. Sustain. Energy Rev.* **2018**, *81*, 1484–1512. [CrossRef]
9. Ren, Y.; Suganthan, P.N.; Srikanth, N. Ensemble methods for wind and solar power forecasting—A state-of-the-art review. *Renew. Sustain. Energy Rev.* **2015**, *50*, 82–91. [CrossRef]
10. Muralitharan, K.; Sakthivel, R.; Vishnuvarthan, R. Neural network based optimization approach for energy demand prediction in smart grid. *Neurocomputing* **2018**, *273*, 199–208. [CrossRef]
11. Alobaidi, M.H.; Chebana, F.; Meguid, M.A. Robust ensemble learning framework for day-ahead forecasting of household based energy consumption. *Appl. Energy* **2018**, *212*, 997–1012. [CrossRef]
12. Mangalova, E.; Agafonov, E. Wind power forecasting using the k-nearest neighbors algorithm. *Int. J. Forecast.* **2014**, *30*, 402–406. [CrossRef]
13. Shang, C.; Wei, P. Enhanced support vector regression based forecast engine to predict solar power output. *Renew. Energy* **2018**, *127*, 269–283. [CrossRef]
14. Tato, J.H.; Brito, M.C. Using smart persistence and random forests to predict photovoltaic energy production. *Energies* **2018**, *12*, 100. [CrossRef]
15. Bracale, A.; Carpinelli, G.; De Falco, P.; Hong, T. Short-term industrial reactive power forecasting. *Int. J. Electr. Power Energy Syst.* **2019**, *107*, 177–185. [CrossRef]
16. Hong, T.; Pinson, P.; Fan, S.; Zareipour, H.; Troccoli, A.; Hyndman, R.J. Probabilistic energy forecasting: Global energy forecasting competition 2014 and beyond. *Int. J. Forecast.* **2016**, *32*, 896–913. [CrossRef]
17. Hong, T. Energy forecasting: Past, present, and future. *Foresight* **2014**, *32*, 43–48.

18. Huang, J.; Perry, M. A semi-empirical approach using gradient boosting and k-nearest neighbors regression for GEFCom2014 probabilistic solar power forecasting. *Int. J. Forecast.* **2016**, *32*, 1081–1086. [CrossRef]
19. Almeida, M.P.; Perpiñán, O.; Narvarte, L. PV power forecast using a nonparametric PV model. *Sola. Energy* **2015**, *115*, 354–368. [CrossRef]
20. Bracale, A.; Carpinelli, G.; De Falco, P. A probabilistic competitive ensemble method for short-term photovoltaic power forecasting. *IEEE Trans. Sustain. Energy* **2017**, *8*, 551–560. [CrossRef]
21. Juban, R.; Ohlsson, H.; Maasoumy, M.; Poirier, L.; Kolter, J.Z. A multiple quantile regression approach to the wind, solar, and price tracks of GEFCom2014. *Int. J. Forecast.* **2016**, *32*, 1094–1102. [CrossRef]
22. Hong, T.; Pinson, P.; Fan, S. Global Energy Forecasting Competition 2012. *Int. J. Forecast.* **2014**, *39*, 357–363. [CrossRef]
23. Wang, Y.; Zhang, N.; Tan, Y.; Hong, T.; Kirschen, D.S.; Kang, C. Combining probabilistic load forecasts. *IEEE Trans. Smart Grid* **2018**, in press. [CrossRef]
24. Golestaneh, F.; Pinson, P.; Gooi, H.B. Very short-term nonparametric probabilistic forecasting of renewable energy generation—With application to solar energy. *IEEE Trans. Power Syst.* **2016**, *31*, 3850–3863. [CrossRef]
25. Gneiting, T.; Raftery, A.E. Strictly proper scoring rules, prediction, and estimation. *J. Am. Stat. Assoc.* **2007**, *102*, 359–378. [CrossRef]
26. Møller, J.K.; Nielsen, H.A.; Madsen, H. Time-adaptive quantile regression. *Comput. Stat. Data Anal.* **2008**, *52*, 1292–1303. [CrossRef]
27. Ziel, F.; Liu, B. Lasso estimation for GEFCom2014 probabilistic electric load forecasting. *Int. J. Forecast.* **2016**, *32*, 1029–1037. [CrossRef]
28. Gaillard, P.; Goude, Y.; Nedellec, R. Additive models and robust aggregation for GEFCom2014 probabilistic electric load and electricity price forecasting. *Int. J. Forecast.* **2016**, *32*, 1038–1050. [CrossRef]
29. R Gbm Package: Generalized Boosted Regression Models. Available online: https://CRAN.R-project.org/package=gbm (accessed on 4 March 2019).
30. Bracale, A.; Carpinelli, G.; De Falco, P.; Rizzo, R.; Russo, A. New advanced method and cost-based indices applied to probabilistic forecasting of photovoltaic generation. *J. Renew. Sustain. Energy* **2016**, *8*, 023505. [CrossRef]

© 2019 by the authors. Licensee MDPI, Basel, Switzerland. This article is an open access article distributed under the terms and conditions of the Creative Commons Attribution (CC BY) license (http://creativecommons.org/licenses/by/4.0/).

Article

Microgrid-Level Energy Management Approach Based on Short-Term Forecasting of Wind Speed and Solar Irradiance

Musaed Alhussein, Syed Irtaza Haider and Khursheed Aurangzeb *

Computer Engineering Department, College of Computer and Information Sciences, King Saud University, Riyadh 11543, Saudi Arabia; musaed@ksu.edu.sa (M.A.); sirtaza@ksu.edu.sa (S.I.H.)
* Correspondence: kaurangzeb@ksu.edu.sa

Received: 15 January 2019; Accepted: 11 April 2019; Published: 19 April 2019

Abstract: Background: The Distributed Energy Resources (DERs) are beneficial in reducing the electricity bills of the end customers in a smart community by enabling them to generate electricity for their own use. In the past, various studies have shown that owing to a lack of awareness and connectivity, end customers cannot fully exploit the benefits of DERs. However, with the tremendous progress in communication technologies, the Internet of Things (IoT), Big Data (BD), machine learning, and deep learning, the potential benefits of DERs can be fully achieved, although a significant issue in forecasting the generated renewable energy is the intermittent nature of these energy resources. The machine learning and deep learning models can be trained using BD gathered over a long period of time to solve this problem. The trained models can be used to predict the generated energy through green energy resources by accurately forecasting the wind speed and solar irradiance. Methods: We propose an efficient approach for microgrid-level energy management in a smart community based on the integration of DERs and the forecasting wind speed and solar irradiance using a deep learning model. A smart community that consists of several smart homes and a microgrid is considered. In addition to the possibility of obtaining energy from the main grid, the microgrid is equipped with DERs in the form of wind turbines and photovoltaic (PV) cells. In this work, we consider several machine learning models as well as persistence and smart persistence models for forecasting of the short-term wind speed and solar irradiance. We then choose the best model as a baseline and compare its performance with our proposed multiheaded convolutional neural network model. Results: Using the data of San Francisco, New York, and Los Vegas from the National Solar Radiation Database (NSRDB) of the National Renewable Energy Laboratory (NREL) as a case study, the results show that our proposed model performed significantly better than the baseline model in forecasting the wind speed and solar irradiance. The results show that for the wind speed prediction, we obtained 44.94%, 46.12%, and 2.25% error reductions in root mean square error (RMSE), mean absolute error (MAE), and symmetric mean absolute percentage error (sMAPE), respectively. In the case of solar irradiance prediction, we obtained 7.68%, 54.29%, and 0.14% error reductions in RMSE, mean bias error (MBE), and sMAPE, respectively. We evaluate the effectiveness of the proposed model on different time horizons and different climates. The results indicate that for wind speed forecast, different climates do not have a significant impact on the performance of the proposed model. However, for solar irradiance forecast, we obtained different error reductions for different climates. This discrepancy is certainly due to the cloud formation processes, which are very different for different sites with different climates. Moreover, a detailed analysis of the generation estimation and electricity bill reduction indicates that the proposed framework will help the smart community to achieve an annual reduction of up to 38% in electricity bills by integrating DERs into the microgrid. Conclusions: The simulation results indicate that our proposed framework is appropriate for approximating the energy generated through DERs and for reducing the electricity bills of a smart community. The proposed framework is not only suitable for different time horizons (up to 4 h ahead) but for different climates.

Keywords: distributed energy resources; energy management; microgrid; deep learning

1. Introduction

The ongoing depletion of fossil fuels, the changing weather, and ecological pollution are some reasons for incorporating DERs into existing power systems. Many advanced countries in the world have directives for energy-providing companies to escalate their energy production from renewable energy sources. In this regard, the government of California established its Renewable Portfolio Standard (RPS) program. In this program, the government signed a bill with utilities to increase the renewable energy production from 20% in 2010 to 33% in 2020 [1].

Table 1 below shows the list of abbreviations used in this paper.

Table 1. List of Abbreviations.

Abbreviation	Full Form
DER	Distributed Energy Resources
IoT	Internet of Things
BD	Big Data
NSRDB	National Solar Radiation Database
NREL	National Renewable Energy Laboratory
RMSE	Root Mean Square Error
MAE	Mean Absolute Error
sMAPE	Symmetric Mean Absolute Percentage Error
MBE	Mean Bias Error
RPS	Renewable Portfolio Standard
ANN	Artificial Neural Network
ELMNN	Extreme Learning Machine Neural Network
GRNN	Generalized Regression Neural Network
SVM	Support Vector Machine
GA	Genetic Algorithm
ANFIS	Adaptive Neuro Fuzzy Inference System
MLP	Multilayer Perceptron
NARX	Nonlinear Autoregressive Recurrent Exogenous Neural Network
CNN	Convolutional Neural Network
LSTM	Long Short Term Memory
EMD	Empirical Mode Decomposition
IMF	Intrinsic Mode Function
RF	Random Forrest
DT	Decision Tree
KNN	K-nearest Neighbors
NWP	Numerical Weather Prediction
CAISO	California Independent System Operator
USA	United States of America
PSM	Physical Solar Model
MH-CNN	Multiheaded Convolutional Neural Network
SG	Smart Grid
ESS	Energy Storage Systems
DNN	Deep Neural Network
MAS	Multi Agent System
PnP	Plug-and-Play
ReLU	Rectified Linear Unit
MSE	Mean Square Error
nRMSE	Normalized Root Mean Square Error
TuO	Time of Use

The energy generation from the DERs is intermittent in nature as it is dependent on naturally varying climate factors, such as wind speed, solar irradiance, and air temperature [2]. These atmospheric variations result in significant changes in the energy generated through DERs, which in turn leads to uncertainty. Consequently, precise and accurate prediction models are crucial for forecasting the generated energy through DERs. These models will be helpful in forecasting the generated energy through DERs, which will be available to the microgrids of smart communities. This will not only help in fulfilling energy requirements but also assist in decreasing energy costs and ensuring the adequate comfort of users in the smart community.

Owing to the intermittent nature of renewable energy resources, the development of a precise and accurate model has become an important factor for increasing the dissemination of DERs in existing power systems. Accurate forecasting of the energy generated through DERs not only helps in the incorporation of renewable energy into power systems but also guarantees good trading performance of renewable energy in the global market [3]. Nevertheless, the forecasting accuracy is heavily dependent on the atmospheric circumstances of the geographical location [4]. Thus, it becomes even more challenging.

There are two main classes of prediction models for forecasting the wind speed and solar irradiance: physical (numerical) models and machine learning (data-driven) models. The main purpose of these models is to forecast wind speed and solar irradiance for a specific location at a selected future time frame. The data-driven models are largely founded on time-series analyses [5]. Their computational complexity is lower than that of the physical models, and they are suitable for short-term prediction. On the other hand, the physical models are based on mathematical equations for relating the dynamics and the physics of the atmosphere, which influences radiation from the sun [1,3,6]. Physical models are usually used for long-term and medium-term forecasting. Consequently, in this work, we selected data-driven models for short-term forecasting owing to their lower complexity and good prediction accuracy.

In the past, traditional statistical approaches have been extensively explored for time-series analyses. Recently, machine learning and deep learning approaches have gained much attention from the research community. Artificial Neural Networks (ANNs) possess exceptional nonlinear mapping and robust generality abilities; thus, these networks can be applied to wind and solar energy forecasting [7]. However, ANN-based models easily fall into local minima and show poor generalization. Moreover, they are well known to over-fit and they have slow convergence rates [8]. In the literature, several other models have been applied, including the Extreme Learning Machine Neural Network (ELMNN) [9], Generalized Regression Neural Network (GRNN) [10], and Support Vector Machine (SVM) [11]. The performance of ELMNN heavily depends on the activation function. If the activation function is not selected appropriately it would result in the generalization degradation phenomenon [12]. Moreover, it is not suitable for applications that require deep extraction of features as it cannot encode more than one layer of abstraction. The main disadvantage of GRNN models is their size and huge computational time [13]. The SVM algorithms have some limitations, such as optimal choice of kernel, computational complexity of the model, and large memory space requirement [14].

Recently, researchers who are applying machine learning models as a core forecasting model have advanced their research with other methods, including weather categorizations [15], parameter or feature selection [16–18], and decorrelation [19]. Some other researchers used hybrid models to enhance the prediction precision. However, the long training time based on increased computational complexity of such models is an issue that needs consideration from the researchers. In a previous study [20], the authors explored an approach for predicting one-day-ahead PV power using neural networks and time-series analysis. The authors in another study [21] implemented and evaluated an optimized prediction model that was based on ANNs and a genetic algorithm (GA). The authors in yet another study [22] explored four architectures (Adaptive Neuro Fuzzy Inference System (ANFIS), Multilayer Perceptron (MLP), GRNN, and Nonlinear Autoregressive Recurrent Exogenous Neural Network (NARX)) for enhancing the prediction precision. They proposed hybrid wavelet-ANN models for solar forecasting at a specific site. However, the model was not tested at different geographical locations to assess the wider potential. Moreover, very limited set of features were used to train the model.

Currently, the convolutional neural network (CNN)-based model is one of the most successful models in deep learning and has been broadly adopted in different applications, including image recognition and classification, object detection, and tracking. However, CNN models have not been extensively explored in time-series analysis. The rapid progress in the computational power of hardware in the last decade has enabled CNN-based models to deeply penetrate various fields. The authors in a previous study [23] proposed a CNN model for interpreting weather data by considering the temporal and spatial associations between the independent parameters for producing local forecasts. They compared the performances of various architectures and stated that the purpose of their exploration was to show that CNN-based models can learn certain patterns of meteorological parameters and relate them to rainfall events. The authors in a previous study [24] also applied a CNN-based model for precipitation prediction.

The authors in a previous study [25] developed a hybrid model by combining long short term memory (LSTM) and CNN models for the prediction of extreme rainfall. The weather parameter was applied as an input to the CNN model, and the outputs of the CNN model were presented as inputs to the LSTM model. In this developed model, the researchers considered the LSTM and CNN models as independent steps. Atmospheric variables, including pressure and temperature, were used as input data. The authors in a previous study [26] developed a framework for the accurate forecasting of short-term wind speed. Their framework was based on hybrid nonlinear/linear models and empirical mode decomposition (EMD). They applied EMD to decompose the wind speed data into residuals and intrinsic mode functions (IMFs). They studied different linear and nonlinear models, including CNN, to analyze the residuals and the IMFs. Among all the hybrid models, EMD-ARMIA-RF performed well for ten-min-ahead forecasting. However, none of the hybrid models performed well 1 h ahead.

In the literature, an approach called benchmarking is mostly used for comparison with the newly developed algorithm [8,27]. The best existing machine learning techniques are selected and evaluated to select the baseline model. The selected baseline model is then used to compare the performance of the newly developed technique. In a previous study [28], authors compared their proposed model for short-term wind speed forecasting with commonly used machine learning algorithms, such as SVM, random forest (RF), and decision tree (DT). We have selected well-known machine learning models, including k-nearest neighbors (KNN), gradient boosting, extra tree regressor, and random forest regressor, for short-term forecasting of wind speed. The best model among them is selected as a baseline model for comparing and evaluating the performance of the proposed model.

Each of the selected machine learning models has its own limitations. For example, KNN algorithm is very sensitive to outliers, as it chose neighbors based on distance criteria. Moreover, it is computationally extensive when the dataset is very large. Gradient boosting models are sensitive to overfitting if the data is noisy. Also, they are harder to tune than other models. One of the weaknesses of random forest and extra tree models when used for regression problems is that the model cannot predict beyond the range in the training data.

Various studies confirm that physical models, such as NWP, are best suited for forecasting more than 4 h to several days [28–35]. These techniques are weak at handling smaller scale phenomenon and are not suitable for short-term forecast horizons [29]. Machine learning methods give the best results for forecast horizons of up to 6 h [30]. The choice of model depends on the forecast horizon. The NWP models generally outperform machine learning models over longer horizons. However, for short-term horizons the time series models have more power [34]. At the intra-hour forecast horizon, NWP is extremely expensive and not practical, especially for the renewable energy sector [36].

In a previous study [37], the authors used smart persistence model as a baseline for the deterministic forecast. In fact, California Independent System Operator (CAISO) uses persistence method in its renewable energy forecasting and dispatching [38]. This method is highly effective in short term prediction, i.e., 1 h ahead. It is often used as a comparison with other advanced methods [39]. In the irradiance forecasting community, numerous works have been devoted recently to the development of models that generate deterministic or point forecasts [34,40–45]. In this work, as we are dealing with short-term forecasting of solar irradiance, we have considered the persistence model and smart persistence model for comparison purposes.

We studied the trends of adaptation of the renewable energy resources in various states of the United States of America (USA). We found that California has made effective policies for the integration of renewable energy resources. In California in 2017, 32% of the electricity was acquired from renewable energy sources, due to which it seems to be well on track to meet its renewable energy targets of 33% and 50% for 2020 and 2030, respectively [46]. Based on the planned effective and concrete policies of the government of California, we have selected San Francisco from the NSRDB of the NREL as a case study in our analysis. The NSRDB uses a physics-based modeling approach, in which the solar radiation data for the entire United States is gridded into segments of 4 km × 4 km using geostationary satellites. The temporal resolution of the data is 30 min [47]. The NSRDB's physics-based, gridded data collection approach is called the Physical Solar Model (PSM). More details about the PSM can be found in a previous study [48].

This paper proposes a multiheaded convolutional neural network (MH-CNN) model for the short-term forecasting of solar irradiance and wind speed to approximate the energy generated through solar panels and wind turbines, respectively. We consider several machine learning models, as well as persistence and smart persistence models for forecasting the short-term solar irradiance and wind speed. We then choose the best model as a baseline and compare its performance with our proposed MH-CNN model. The comparison is based on evaluation metrics, including the root mean square error (RMSE), mean absolute error (MAE), and symmetric mean absolute percentage error (sMAPE). Using the NSRDB of the NREL data of San Francisco as a case study, the results show that our proposed model outperforms all other models in forecasting the wind speed and solar irradiance. The obtained results indicate that our proposed framework for microgrid-level energy management is appropriate for approximating the renewable energy and for reducing the electricity bills of a smart community. The main contributions of our work are as follows:

- We formulated a solar irradiance and wind speed prediction problem for approximating the generated energy through solar panels and wind turbines.
- We evaluated the performance of various machine learning models, as well as persistence and smart persistence models, for selecting a baseline model.
- We proposed an MH-CNN model for the short-term forecasting of solar irradiance and wind speed.
- We evaluated the effectiveness of proposed model on different time horizons (up to 4 h).
- We evaluated the effectiveness of proposed model on different climates.
- We proposed a framework for microgrid-level energy management for reducing the electricity bills of a smart community.

The remainder of the paper is organized as follows. The related work from the literature is reviewed and presented in Section 2. The proposed framework is elaborated in Section 3. A performance evaluation of the various considered models is provided in Section 4. Results and discussions are provided in Section 5. The contributions of the paper are discussed in Section 6.

2. Related Work

Researchers from academia and industry have explored methods and technologies for tackling the problems of global energy crises. In this part of the manuscript, we present recent research work on global energy crises and explored solutions. The future Smart Grid (SG) will be composed of the latest technologies and will significantly improve the existing power grids. The possibility of two-way information flow and interoperability between smart homes provides a chance to optimize the power consumption of the end users and simultaneously improve the operation of the SG [49–52]. The increasing diffusion of renewable energy in power systems has given rise to the concept of microgrids, which will probably play a substantial role in the development of SGs [53,54]. It is anticipated that the network of microgrids will result in the formation of an SG [54]. The microgrid is composed of DERs, power loads, and Energy Storage Systems (ESS) [55,56].

Typically, DERs, such as wind turbines and solar panels, are among the useful energy resources for solving energy shortfalls. These resources also help in decreasing the effects of carbon emissions in the modern world. By incorporating DERs in power systems, consumers will be able to achieve their power requirements by generating green energy, which in turn will lead to electricity bill reductions. In the last decade, many researchers focused their efforts on solving the challenges of DERs—their integration into the SG, intermittent nature, the optimal power flow, etc. One of the important issues with the energy generated through DERs is the intermittent nature of these power-generating sources. Many researchers have dedicated their efforts to mitigating these issues [57,58]. The development of an accurate prediction model for forecasting the wind energy and solar energy is desirable. However, the generated energy from DERs is heavily reliant on the accuracy of the weather prediction model.

The accuracy of the weather prediction model is reliant on different atmospheric phenomena, such as pressure, temperature, wind speed, and humidity. The enormously random variations of weather conditions lead to difficulties in the accuracy of the prediction [58]. Fortunately, different parameters of the weather can be predicted with significant accuracy by developing any of the latest models, including the ANN, Deep Neural Network (DNN), and LSTM [59–61].

The authors investigated the integration of DERs in power systems in a previous study [62]. They suggested dealing with the uncertainty of DERs by virtualization, and validated their method by performing real-time experiments. The authors in another study [63] investigated a prediction model for approximating the quantity of solar energy generation. Their prediction model was composed of a wavelet transform and a neural network. They used RMSE and MAE to evaluate their developed model. A comparison of their obtained results with existing promising results proved that their developed model achieved good performance.

Recently, in a previous study [64], we proposed a short-term load prediction technique based on support vector quantile regression. In this study, we compared three kernel functions: Gaussian kernel, linear kernel, and polynomial kernel. The predicted precision of the power load was approximated using data sets from Singapore. We achieved better results compared to those from Support Vector Regression and the Firefly Algorithm. Power systems in today's world are being transformed into distributed energy resources. The integration of DERs in existing power systems leads to energy management problems because these energy resources produce power in nondeterministic manners. The well-known "duck curve" problem arises in the off-peak hours because of the overgeneration from DERs that causes generator units to be underloaded [65]. The underloading of a generator impacts the individual components of a power system and the overall system performance because of the mismatch between generation and demand.

Recently, in a previous study [66], we considered the radial structure of a distribution grid and applied commonly used configuration topology for the integration of DERs and ESSs in power systems. Furthermore, we addressed a multilevel Multi Agent System (MAS) optimization framework for the co-scheduling of demand and supply resources. The MAS structure permits Plug-and-Play (PnP) capabilities and flexible control of DERs for load balancing. During both off-peak and peak hours, the PnP algorithm deactivates or activates the ESS to rectify demand and supply mismatches. The ESS stocks the excess energy from DERs and uses it to meet the energy demand at a later time. Our main objective has been to reduce electricity bills without compromising user comfort during peak hours. Our simulation results proved that our developed MAS helped in balancing the load while maintaining adequate user comfort.

In the current work, our aim is to develop an MH-CNN model for the short-term forecasting of wind speed and solar irradiance. The forecasted wind speed and solar irradiance will be used for approximating the generated power through wind turbines and solar panels. We performed extensive simulations to prove the improved performance of the proposed strategy. Moreover, we proposed a framework for microgrid-level energy management for reducing the electricity bills of the smart community.

3. Proposed System Model

In this section, the proposed system model is explained. It is always beneficial to reduce the electricity bill of the users without affecting their comfort. Integrating DERs in the power system helps to reduce electricity bills, increase user comfort, and fulfill energy requirements. Sometimes the total electricity generated by DERs during off-peak hours exceeds the demand of the consumer, which results in a generation-demand imbalance. The consequence of a generation-demand imbalance is the basis of the "duck curve" problem [65]. Temporarily, excessive electricity generation from DERs lessens the power load on the grid generators. In this situation, the excess power generated by the DERs may be harmful to the generator and motors. Thus, there is a need to develop efficient machine and deep learning models to accurately predict short-term renewable energy generation. Based on these models, efficient energy management frameworks need to be explored.

The proposed microgrid-level energy management framework is presented in Figure 1, where a smart meter, ESS, and DERs are integrated. As shown in Figure 1, an ESS is integrated in the proposed system to mitigate the influence of the duck curve problem, which we recently targeted in a previous study [66]. The smart meters are used for two-way communication in addition to many other advanced features. The DERs in the form of wind turbines and solar panels are used to generate the renewable energy to ensure the required user comfort and to reduce electricity bills. The ESSs are used for storing the excess generated energy at any time. This excess energy can then be used at a later time. In addition to DERs, the microgrid has access to power from the main grid, as the nature of the DER is intermittent and may produce very low energy on certain days and at certain times.

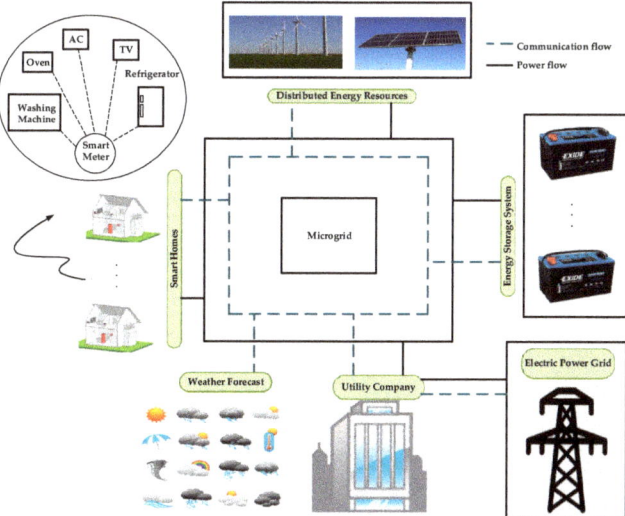

Figure 1. Framework for microgrid-level energy management.

The architecture of the proposed MH-CNN model is shown in Figure 2. We used the same model for the short-term forecasting of both wind speed and solar irradiance. Meteorological parameters, such as temperature, pressure, and wind speed, as well as cyclic parameters, such as season, month, day of the year, and hour over the past day, are passed to both the wind speed and solar irradiance forecasting models. Moreover, we incorporate the past day's lag of wind speed and solar irradiance as lag features in the wind speed and solar irradiance forecasting models, respectively. The data preparation steps are presented in Figure 3.

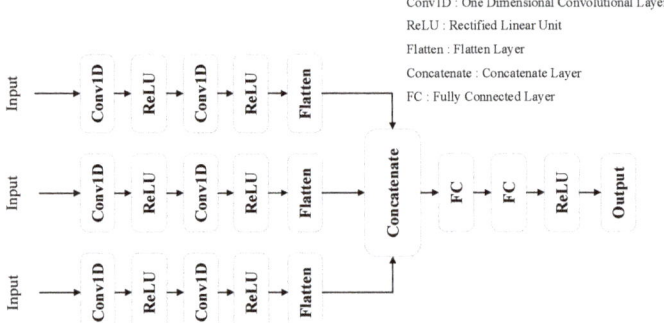

Figure 2. Proposed MH-CNN model.

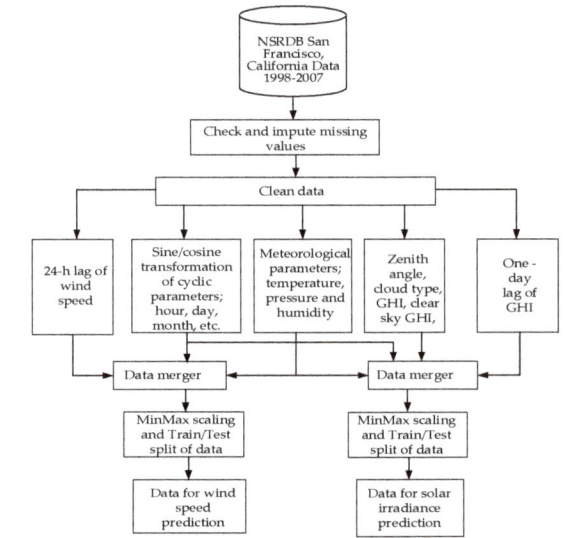

Figure 3. Preprocessing steps for cleaning and separating data into train test sets.

The same input is passed to three 1D CNNs. Each CNN has the same filter size but different kernel sizes. All three sub-CNN models extract features by looking at the input data from different aspects owing to the different kernel sizes. For our model, we used a Rectified Linear Unit (ReLU) as an activation function, as it does not encounter the gradient vanishing problem and performed best in the case study data. The CNN part consists of two 1D convolution layers. In the second convolution layer, we halved the filter size and doubled the kernel size to reduce the dimensionality and enhance the feature selection domain, respectively. The output of the second convolution layer (after applying the ReLU activation function) for each sub-CNN model is flattened and concatenated as a single feature vector. The feature vector then goes through the fully connected architecture and the ReLU activation function to produce the output, as shown in Figure 2.

The data preprocessing steps are shown in Figure 3. Initially, the missing values are determined and are replaced with the values from the same time on the previous day. If the value from the same time of the previous day is also missing, then the missing data is imputed by using the value of the same time of the last previous day with available data. Then, further processing is performed on the clean data in three different ways. The sine and cosine transformations of cyclic parameters, such as hour of day, day of the year, month of the year, season of the year, and wind direction, are determined. We used binary encoding to encode the categorical feature named "cloud type". This feature was obtained by the NREL from the pathfinder atmospheres extended (PATMOS-X) model. Meteorological parameters, including temperature, pressure, and wind speed, are separated. The one-day time lags of the wind speed and solar irradiance are arranged as separate features for the wind speed and solar irradiance models, respectively. These features are merged, and then normalization is performed, i.e., the range of each input vector is restricted to (0, 1). The scaled data are then separated into train and test data sets for training and evaluating the proposed model, respectively.

3.1. Convolutional Neural Networks

In this section, we describe the layers associated with the implementation of our forecasting model, including 1-D convolution, ReLU, dropout, and fully-connected layers.

3.1.1. The 1-D convolution

The convolutional layer is the most important building block of any CNN. This layer is regarded as a set of learnable filters that consists of many convolution operations. The parameters of every convolution operation are optimized by a back propagation algorithm. Each filter in a specific convolution layer has the same receptive field. An example of 1D convolution is shown in Figure 4. The weights associated with kernel size of 3 are {w1, w2, w3}. These weights are shared by the input layer {i1, i2, i3, i4, i5}. The feature map will be obtained by the convolution between the weights and inputs. In this example, the feature f2 is obtained by f2 = w1 × i2 + w2 × i3 + w3 × i4.

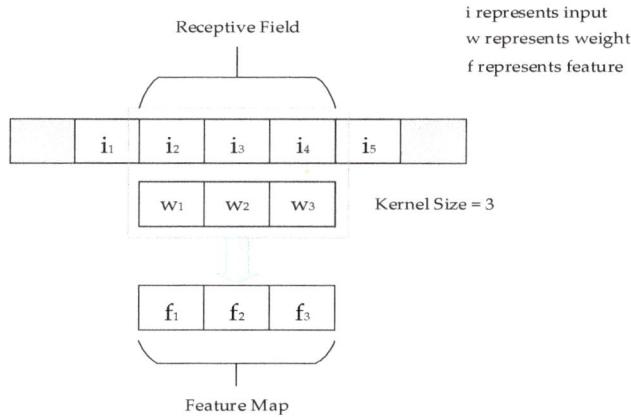

Figure 4. Example of 1-D convolution operation.

3.1.2. *ReLU*

The activation functions are used to enhance the ability of models to learn complex structures. ReLU has been widely adopted by various researchers to make the network more trainable. It works by thresholding values at 0, i.e., f (z) = max (0, z).

3.1.3. Dropout

The dropout technology provides an easy way to overcome the overfitting problem while designing the deep learning model. This method involves the random selection of neurons and disabling them during training. The output values of these randomly disabled neurons are zero.

3.1.4. Fully-Connected Layer

The fully connected layer exhibits the nonlinear mapping from the input to the output, by using bias and an activation function. These layers are usually applied towards the end of the network. We use the flatten layer after the convolution layers, as this layer expects 1-D data.

3.2. Proposed Model

The details of our proposed MH-CNN model are shown in Figure 5. The number of features for wind speed and solar irradiance short-term forecasting are 62 and 47, respectively. For wind speed forecasting, there are 48 instances per day, as the data are recorded every 30 min. However, for solar irradiance forecasting, we considered the data from 5:30 to 19:00; hence, there are 28 instances per day. We used 64 filters in the first convolution layer with kernel sizes of 3, 5, and 7 for each head of the MH-CNN. Similarly, we used 32 filters in the second convolution layer with kernel sizes of 5, 7, and 9. We used a dropout value of 0.5 before applying the flattening layer. After concatenating all of the features into a single feature vector, a fully connected layer was applied with 16 neurons and the ReLU as a nonlinear activation function. The parameter settings of the proposed model are listed in Table 2. The prediction of both wind speed and solar irradiance concerns half-hour-ahead prognosis. The proposed model forecasts the next half hour value using the values of the previous day as inputs.

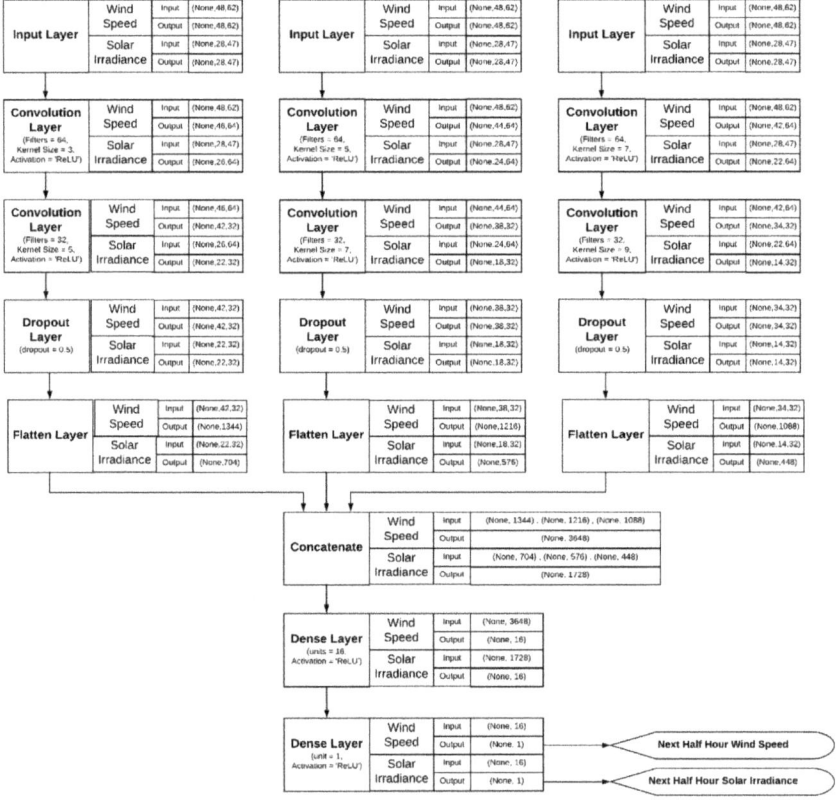

Figure 5. Structure of proposed MH-CNN model.

Table 2. Parameter settings of proposed MH-CNN model.

Parameter	Setting
Optimizer	Adam
Training Stop Strategy	Early Stopping Criteria
Loss Function	Mean Square Error (MSE)
Learning Rate	{0.0001}
Batch size	{128}
Epoch	{200}

The training flow of the proposed model is shown in Figure 6. The training data are split into 90% training data and 10% validation data. The validation loss is based on the Mean Square Error (MSE) value. If the validation loss does not decrease for two consecutive epochs, then the learning rate is reduced by a factor of 0.85. The minimum value of the learning rate is set to be 1×10^{-6}. If the validation loss is decreasing, then the model is saved with the updated weights. To avoid overfitting of the model during the training process, if the validation loss is not decreasing for 10 consecutive epochs, then early stopping callback is applied, and the last-saved best model is loaded for forecasting and performance evaluation. Otherwise, the training process continues until the maximal number of epochs is completed.

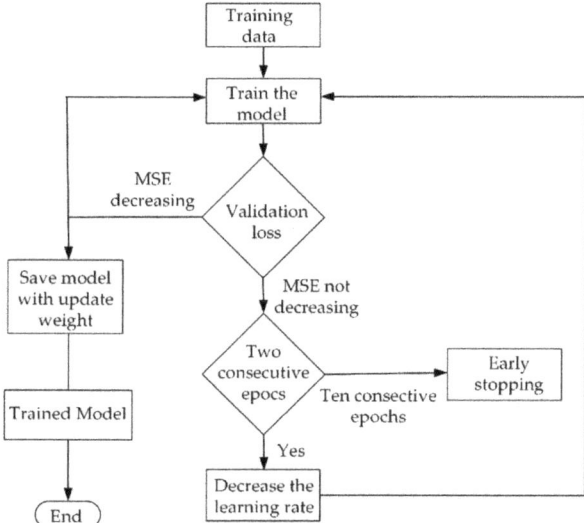

Figure 6. Training flow of proposed model.

The performance of the trained model is evaluated based on various metrics, including RMSE, MAE, and MBE. The mathematical calculation methods of these performance matrices with their equations are shown in Equations (1) to (3).

$$RMSE = \frac{1}{\sqrt{N}} \sqrt{\sum_{m=1}^{N} (a_m - p_m)^2} \tag{1}$$

$$MAE = \frac{1}{N} \sum_{m=1}^{N} |a_m - p_m| \tag{2}$$

$$MBE = \frac{1}{N}\sum_{m=1}^{N}(a_m - p_m) \tag{3}$$

where a_m is the actual value, and p_m is the predicted value. The RMSE, MAE, and MBE represents model prediction error in units of the target variable. The RMSE gives a relatively high weightage to the outliers compared to MAE, as the residual is squared before averaging. The MAE is a linear score where all the individual differences are weighted equally. The MBE indicates the degree to which the observations are "over" or "under" forecasted by the prediction model. The smaller RMSE, MAE, and MBE denote the good performance of a forecasting model. MAPE is another standard metric for evaluating the performance of forecasting algorithms. Problems in its use can occur when a_m is zero or very small. As an alternative, we used sMAPE, as shown in Equation (4):

$$sMAPE = \frac{2}{N}\sum_{m=1}^{N}\frac{|a_m - p_m|}{|a_m| + |p_m|} \tag{4}$$

4. Performance Evaluation of Proposed Model

In this study, weather data from 1998 to 2007 for San Francisco, California, are used. The data were retrieved from the NSRDB of the NREL [67]. The first nine years of data are used to train the model, and the data for the last year (2007) are used to test the performance of the trained model.

4.1. Short-Term Forecasting Analysis of Wind Speed

We selected KNN, gradient boosting, extra tree regressor, and random forest regressor as our machine learning models. The input to all these models is the complete set of features mentioned in Figure 3. We used the default values of the parameters for our baseline model comparison. We used MSE as the loss function for all machine learning models in this study.

We also considered the persistence model for the short-term forecasting of wind speed. The persistence model assumes that wind data at a certain future time (the next half hour, in our case) will be the same as when the forecast was made. In this study, for the persistence model, we assumed that the wind data in the next half hour will be the same as that of the current time.

For our baseline model comparison, the parameters of various machine learning algorithms are taken from a previous study [8] and shown in Table 3.

Table 3. Parameters of various machine learning algorithms [8].

No.	Model	Parameters
1	KNN	No. of neighbors, $n = 5$; weight function = uniform; distance metric = Euclidian
2	Gradient Boosting	No. of estimators = 100; Maximum depth = 75; minimum samples split = 4; minimum sample leaf = 4,
3	Extra Tree	No. of trees = 100; maximum depth of tree = 100; min samples split = 4; min sample leaf = 4
4	Random Forrest	No. of trees = 125; maximum depth of tree = 100; min samples split = 4; min sample leaf = 4

All trained machine learning models were evaluated on the same test data. We selected three standard evaluation metrics for comparing the performance of these models: RMSE, MAE, and sMAPE. Based on the test data set of one year, i.e., 2007, the seasonal average values (three-months-average values of RMSE, MAE, and sMAPE, with spring season defined as March, April, and May) are calculated. The seasonal variation of RMSE, MAE, and sMAPE are shown in Figures 7–9, respectively. It is clear from these figures that the random forest method outperforms the persistence model, and therefore serves as the baseline model. The RMSE, MAE, and sMAPE of our proposed model are the lowest among the evaluated models.

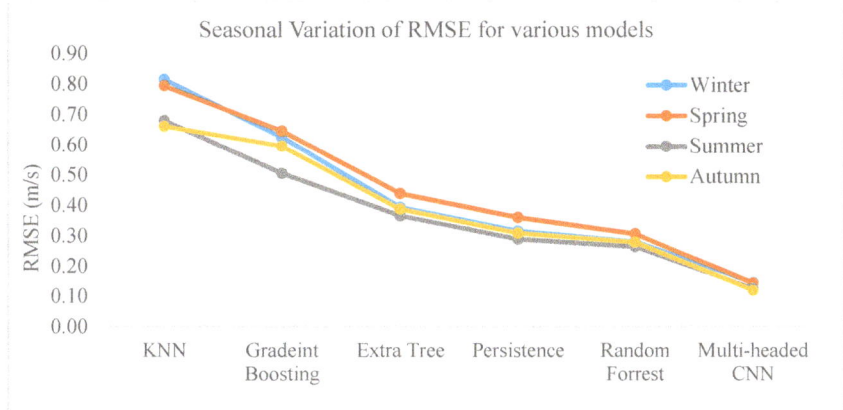

Figure 7. Seasonal variation of RMSE for forecasting wind speed.

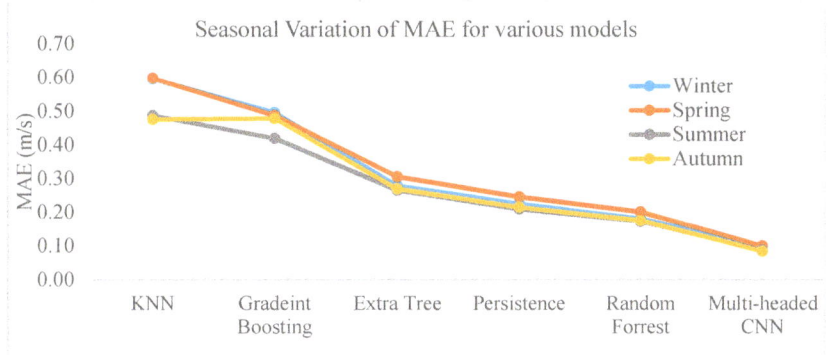

Figure 8. Seasonal variation of MAE for forecasting wind speed.

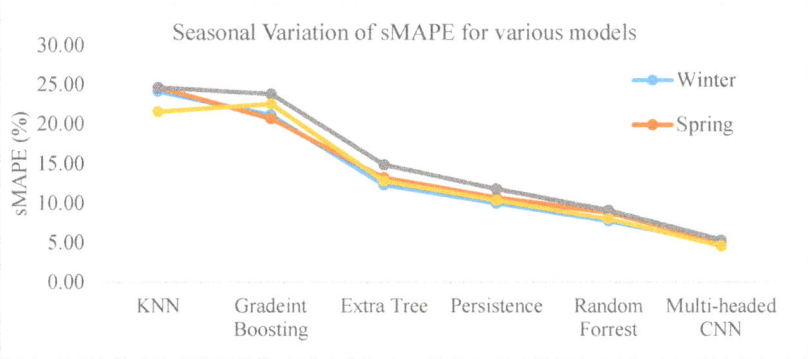

Figure 9. Seasonal variation of sMAPE for forecasting wind speed.

We have selected a random day from the test data to demonstrate the comparison of various machine learning models with the proposed model. Detailed comparison results of the wind speed prediction for all of the evaluated models are presented in Figure 10. The bold blue line represents the actual wind speed, whereas the bold black line represents the forecast by the proposed model. A careful analysis of this figure reveals that the forecast results of the KNN and gradient boosting algorithms barely coincide with the actual wind speed. The wind speed predicted by our proposed model is quite close to the actual wind speed. The forecasting ability of the proposed model is also verified in this experiment. The independent axis in this figure shows 48 values because of the 30 min sampling interval of the measured data.

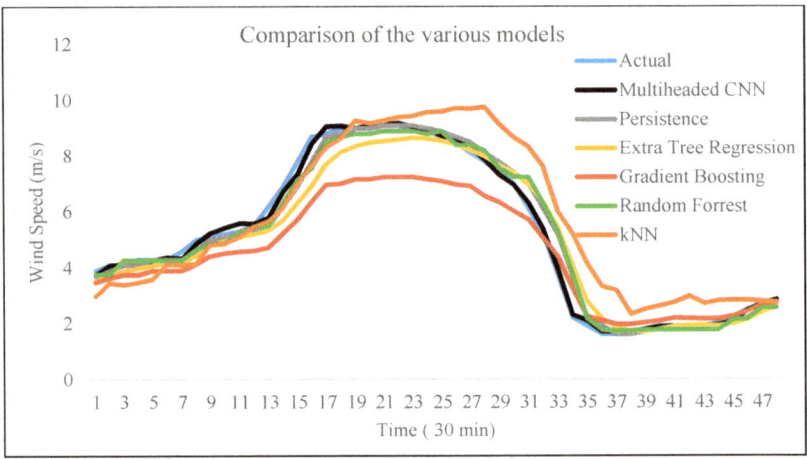

Figure 10. Comparison of wind speed prediction for various models.

In Figure 11, a scatter plot of the predicted and actual wind speed values for the complete test data is presented. The coefficient of determination value is 0.9948, which confirms the strong, positive, linear association between the predicted and actual wind speeds. Furthermore, the coefficient of determination shows that the proposed model is able to explain 99.48% of the variation of the actual data. This indicates the very good forecasting ability of the proposed model.

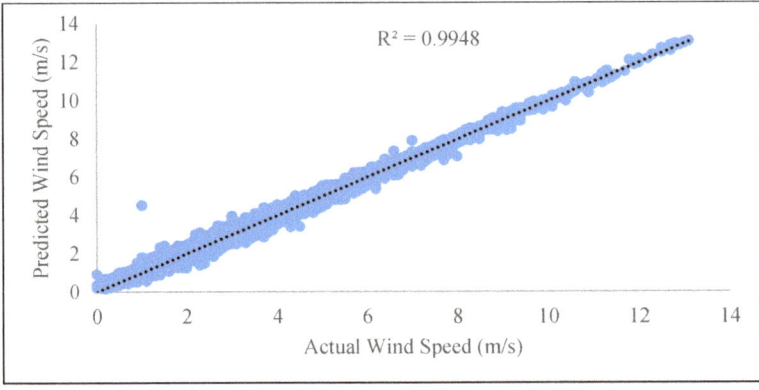

Figure 11. Actual vs. predicted wind speed for test data.

Previous results (Figures 7–9) showed the evaluation performance of various models for seasonal trends. Furthermore, we evaluated the performance of the selected models on the complete test data. These results are listed in Tables 4 and 5.

Table 4. Comparison of various models based on test data.

Model	RMSE (m/s)	MAE (m/s)	sMAPE (%)	R^2
Persistence	0.2819	0.1820	8.41	0.9772
k-Nearest Neighbor	0.7373	0.5385	23.71	0.8406
Random Forest	0.2432	0.1702	7.17	0.9809
Gradient Boosting	0.5924	0.4694	22.01	0.8976
Extra Tree	0.3959	0.2792	13.28	0.9527

Table 5. Comparison of baseline model with proposed model.

Model	RMSE (m/s)	MAE (m/s)	sMAPE (%)
Random Forest	0.2432	0.1702	7.17
Proposed Model	0.1339	0.0917	4.92
Error Reduction (%)	44.94	46.12	2.25

Table 4 shows a comparison of the various models on the basis of the evaluation metrics for the selection of a baseline model. The results indicate that the KNN and gradient boosting algorithms perform poorly on the complete test data. Usually, it is difficult to outperform the persistence model for short-term forecasting. We can see that the random forest method outperformed the persistence model. Therefore, we selected random forest as a baseline model for comparison with our proposed model.

The comparison of our proposed model with the baseline model is presented in Table 5. It is clear from this table that the proposed MH-CNN model resulted in much lower (better) evaluation metrics than the baseline model. The percentages of error reductions achieved by the proposed model for RMSE, MAE, and sMAPE are 44.94, 46.12, and 2.25, respectively.

4.2. Short-Term Forecasting Analysis of Solar Irradiance

In solar irradiance forecasting, the persistence model usually serves as a baseline model for short-term forecasting. A simple persistence model [36] is shown in Equation (5):

$$GHI_{persistence}(t + \Delta t) = GHI(t) \qquad (5)$$

where $GHI(t)$ is the current global horizontal irradiance (GHI) at the surface. (The terms "solar irradiance" and "GHI" are used interchangeably throughout the manuscript.)

We selected another model with which to compare our proposed model: a variant of the persistence model, called the smart persistence model [36]. It is defined as

$$GHI_{smart\ persistence}(t + \Delta t) = k_t(t) \times GHI_{clear\ sky}(t + \Delta t) \qquad (6)$$

where $k_t(t)$ is the clear-sky index correction factor, defined as

$$k_t(t) = \frac{GHI(t)}{GHI_{clear\ sky}(t)} \qquad (7)$$

Based on the test data of solar irradiance for the year 2007, Figures 12 and 13 show the seasonal average variations for the persistence, smart persistence, and proposed models in terms of RMSE and sMAPE, respectively. It is clear from these figures that RMSE and sMAPE of our proposed model are lower than those of the persistence and smart persistence models. This shows that our proposed model is suitable for the short-term forecasting of solar irradiance.

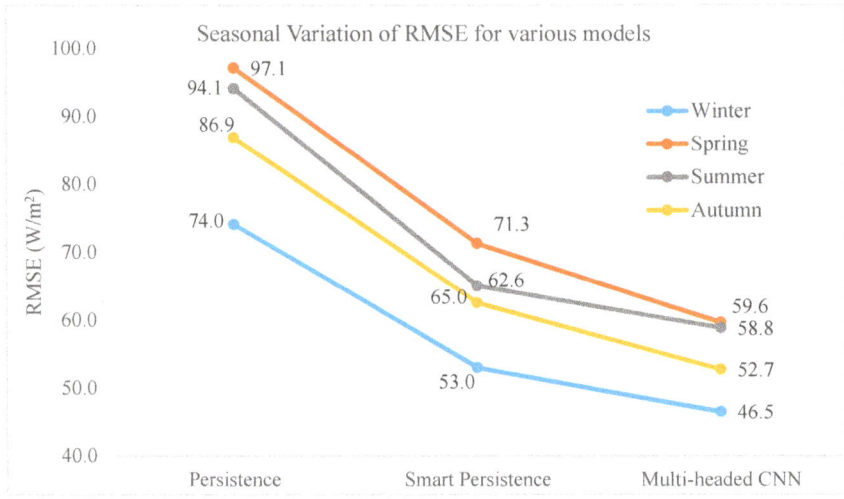

Figure 12. Seasonal variation of RMSE for forecasting solar irradiance.

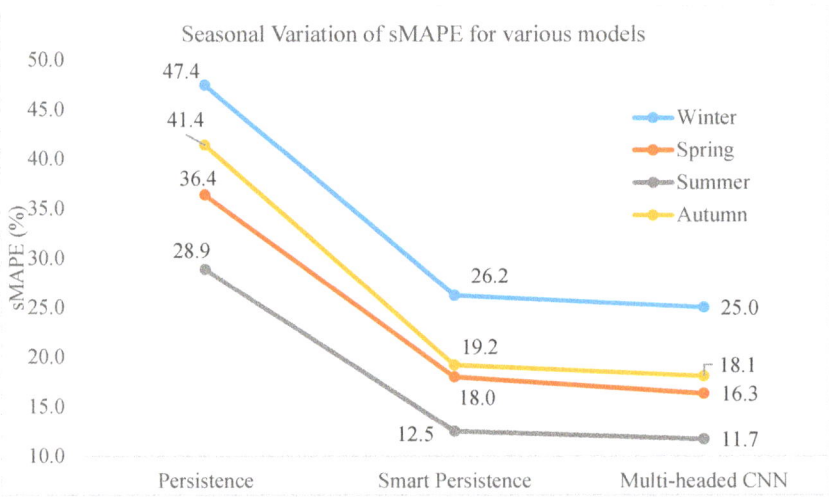

Figure 13. Seasonal variation of sMAPE for forecasting solar irradiance.

A comparison of the actual GHI and predicted solar irradiance using the proposed model and the smart persistence model is shown in Figure 14a. We selected a random day from the test data to demonstrate the comparison of the smart persistence model with the proposed model. The prediction accuracies of the proposed model and smart persistent model are shown in Figures 14b and 14c, respectively. The coefficient of determination of the proposed model is reasonably high compared to that of the smart persistence model, which shows that our proposed model can be successfully applied to predict the solar irradiance. Later, we used the predicted solar irradiance to approximate the generated solar energy.

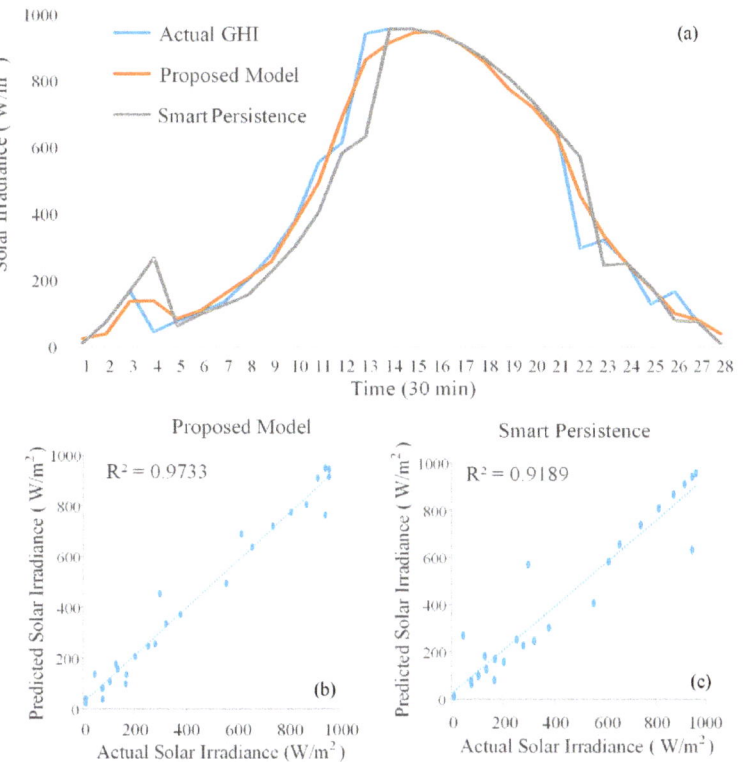

Figure 14. (**a**) Comparison of actual GHI and predicted solar irradiance by smart persistence and proposed model, (**b**) prediction accuracy of proposed model, and (**c**) prediction accuracy of smart persistence.

A comparison of the persistence and smart persistence models for the selection of the best model is presented in Table 6. As seen from the results, the smart persistence model is best as a baseline model for comparison with the proposed model.

Table 6. Comparison of persistence and smart persistence models based on test data.

Model	RMSE (W/m^2)	MBE (W/m^2)	sMAPE (%)	R^2
Persistence	83.21	0.00	38.52	0.9308
Smart Persistence	59.22	3.85	18.08	0.9648

A comparison of the proposed model with the smart persistence model is presented in Table 7. It is clear from the results of this table that the proposed model produced better results than the smart persistence model. We achieved 7.68%, 54.29%, and 0.14% error reductions in the RMSE, MBE, and sMAPE, respectively.

Table 7. Comparison of proposed and smart persistence models based on test data.

Model	RMSE (W/m^2)	MBE (W/m^2)	sMAPE (%)
Smart Persistence	59.22	3.85	18.08
Proposed Model	54.67	1.76	17.94
Error Reduction (%)	7.68	54.29	0.14

5. Results and Discussions

In this section, we evaluated the effectiveness of the proposed model for different forecasting horizons and different climates. Then, we performed a comprehensive bill reduction analysis by estimating the generated power from renewable resources using the data of San Francisco as a case study.

5.1. Evaluation of Proposed Model for Different Time Horizons

In the previous section, we explored the effectiveness of the proposed model for a single-step forecast, i.e., predicting the observation at the next time stamp. To illustrate the effectiveness of the proposed model for multi-step forecasting, i.e., different time horizons, we considered the same data used in Section 4. The output of the proposed model was reshaped according to the forecasting time horizon. For example, when the time horizon was set to 4 h ahead, then the model was evaluated such that for each instance of test data, the model will predict the next eight values in one-shot.

Table 8 shows the seasonal RMSE variation of the proposed model for wind speed and solar irradiance forecasting of different time horizons.

Table 8. Effectiveness of proposed model in different time horizons.

Season	Wind Speed RMSE (m/s)			Solar Irradiance RMSE (W/m^2)		
	Half Hour Ahead	1 h Ahead	4 h Ahead	Half Hour Ahead	1 h Ahead	4 h Ahead
Summer	0.13	0.21	0.50	58.81	65.40	90.22
Autumn	0.12	0.22	0.41	52.77	60.66	85.72
Winter	0.15	0.24	0.48	46.49	53.65	72.45
Spring	0.14	0.24	0.58	59.60	66.80	87.87

5.2. Evaluation of Proposed Model for Different Climates

In Section 4, we tested the proposed model on San Francisco (Latitude: 37.77, Longitude: −122.42), which has a warm summer Mediterranean climate. In order to demonstrate the effectiveness of the proposed model in a different climate, we selected New York (Latitude: 40.73, Longitude: −74.02), which has a humid subtropical climate, and Las Vegas (Latitude: 33.61, Longitude: −114.58), which has a hot desert climate. In this experiment, data from 1998 to 2007 for New York and Las Vegas is used. The data were retrieved from the NSRDB of the NREL [67]. We prepared the training and test data according to Figure 3 for both sites. The first nine years of data (1998–2006) was used to train the model and the last one year of data (2007) was used to test the performance of the trained model.

To fairly compare the RMSE across different sites, normalized root mean square error (nRMSE) is computed as,

$$nRMSE = \frac{RMSE}{\bar{y}} \qquad (8)$$

where $\bar{y}$ is the mean of the actual values. Table 9 shows the seasonal variation of the proposed model for short-term forecasting of wind speed and solar irradiance for different climates.

Table 9. Effectiveness of proposed model in different climates.

Season	Wind Speed nRMSE			Solar Irradiance nRMSE		
	San Francisco	New York	Las Vegas	San Francisco	New York	Las Vegas
Summer	0.055	0.035	0.0416	0.1060	0.1903	0.0918
Autumn	0.046	0.029	0.0531	0.1427	0.1950	0.0750
Winter	0.049	0.032	0.0430	0.1691	0.2251	0.1181
Spring	0.049	0.040	0.0489	0.1234	0.2167	0.0903

As seen in the table, for each season, there is a small discrepancy between the wind speed nRMSE of various sites with different climates. This result indicates that our proposed model is capable of forecasting the short-term wind speed for different climates during various seasons with high accuracy.

For short-term forecasting of solar irradiance, there is an almost 9% difference between the best predictor (summer season) for San Francisco and New York. This discrepancy is certainly due to the cloud formation processes, which are very different in these two sites. The two sites experience different sky conditions during the year. Sites such as San Francisco and Las Vegas exhibit stable sky conditions during the summer. However, New York witnesses occasional thunderstorms with heavy rain in summer, and tornadoes are not uncommon.

There is an almost 5.6% difference between the worst predictor (winter season) of San Francisco and New York. In San Francisco and New York, the sky is mostly cloudy, around 55% and 53% of the time in winter, respectively. The proposed model performance is worst in winter, since the sky coverage is highly variable. In Las Vegas, there is a significant seasonal variation in the cloud coverage over the course of the year. For a hot desert climate, such as Las Vegas, the seasonal performance of the model is reasonably suitable for solar irradiance forecasting.

The result of the solar irradiance forecast indicates that the proposed model is well-suited for a hot desert climate, as well as a Mediterranean climate. Moreover, it can also be used for a humid subtropical climate.

5.3. Generation Estimation and Bill Reduction Analysis: San Francisco as a Case Study

We considered a smart community consisting of 80 homes as the consumers of the electricity. For simulation purposes, it was assumed that the smart community has a microgrid that is equipped with wind turbines and solar panels, in addition to having access to the power from the main grid. At any time, the energy generated by the wind turbine and solar panels is provided to the users through the microgrid, and the excess generated energy is stored in the ESS for later use. In addition, the deficit energy at any time is purchased from the commercial grid to satisfy the energy demands of the users in the smart community.

In this work, a Time of Use (ToU) pricing model is applied to determine the price of the consumed electricity [68], which is shown in Figure 15. A 24-h time period is considered and is denoted by T. This is divided into 1-h subintervals indicated by t.

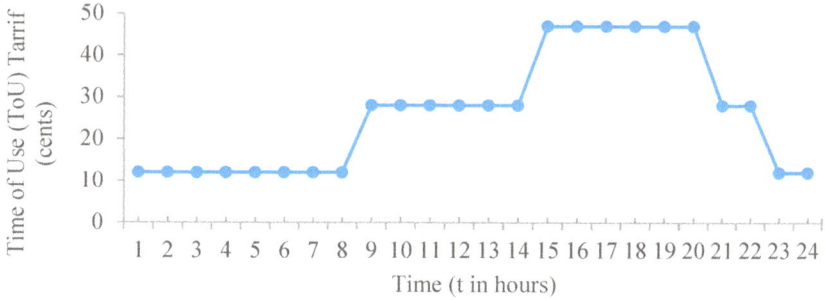

Figure 15. Applied Time of Use (ToU) electricity tariff [68].

For a randomly selected day, the proposed model is applied to predict the wind speed for 24 h. The predicted wind speed is then used to approximate the generated wind energy based on Equation (8), which we also used in our recent work [69]. The predicted wind speed and generated wind power are presented in Figure 16. It is clear from the figure that with an increase in the wind speed, the power generated by the wind turbine also increases. The power generation of the wind turbine is approximated by implementing Equation (8) in MATLAB. For simulation purposes, we used a single wind turbine of 30 kW [70]. As shown in the figure, when the wind speed is equal to, or greater than, the rated wind speed of the selected wind turbine, the output power is the maximum attainable, which is the rated maximum power.

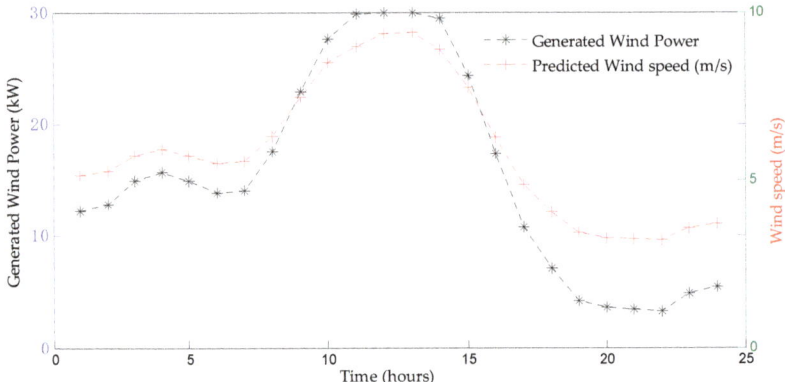

Figure 16. Predicted wind speed and approximated wind power.

The predicted half-hour wind speed for three days and the associated generated wind power are shown in Figure 17. The fluctuations in the predicted wind speed at different times during the 72-h period are evident from the figure. As shown, when the wind speed is lower than the cut-in speed of the wind turbine, the generated power is zero.

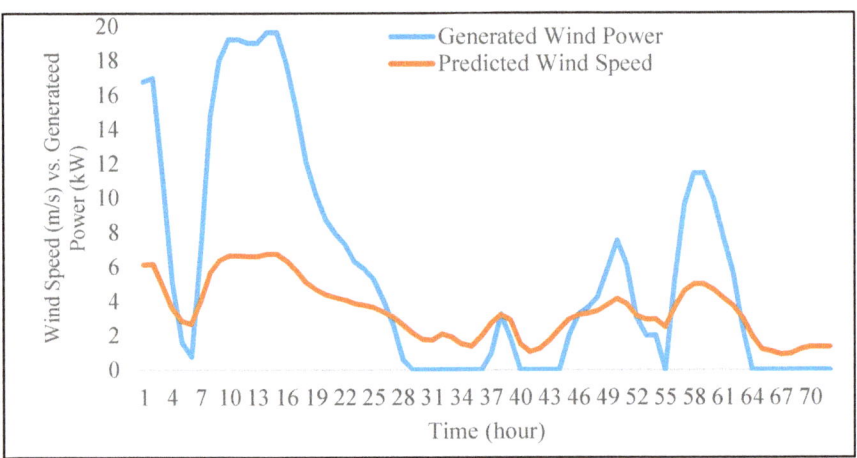

Figure 17. Predicted wind speed for three days and associated power generated by wind turbine.

In Equation (9), the power generated by the wind turbine is represented by P_t^{wt}, and C_p is the power coefficient. It also depends on air density ρ, area swept by rotor blades A, and wind speed V_t^{wt}. The wind turbine triggers are based on the cut-in and cut-out speeds. The association between the output power and the wind speed of the wind turbine is based on Equation (10) from a previous study [71]. In Equation (10), P_{out} is the output power, P_R is the rated power, V_t^{wt} is the wind speed at time t, V_{ci} is the cut-in wind speed, and V_{co} is the cut-out wind speed. The technical specifications of the selected wind turbine are shown in Table 10.

$$P_t^{wt} = \frac{1}{2} C_p \, \rho \, A \, (V_t^{wt})^3 \qquad (9)$$

$$P_{out} = \begin{cases} 0, & 0 <= V_t^{wt} < V_{ci} \\ P_R \frac{V_t^{wt} - V_{ci}}{V_R - V_{ci}}, & V_{ci} <= V_t^{wt} < V_R \\ P_R, & V_R <= V_t^{wt} < V_{co} \\ 0, & V_t^{wt} >= V_{co} \end{cases} \quad (10)$$

Table 10. Technical specifications of the selected wind turbine.

Item	Symbol	Description/Value
Turbine Model	-	Aeolos-H 30kW
Rotor Diameter (m)	-	15.6
Rated Power (kW)	P_R	30.0
Cut-in Wind Speed (m/s)	V_{ci}	2.5
Rated Wind Speed (m/s)	V_R	9.0
Cut-out Wind Speed (m/s)	V_{co}	25.0

Figure 18 reveals the association between the power generated by the solar panel and the solar irradiance. We can observe that with an increase in the solar irradiance, the power generated by the solar panel also increases. The solar panel temperature data are taken from a previous study [72]. The solar irradiance is predicted using our proposed model. The power generation by the solar panel is approximated by implementing Equation (11) from our previous work [69] in MATLAB.

$$P_t^{pv} = \eta^{pv} A^{pv} \ Irr \ \left(1 - \frac{1}{200}(T_t - 25)\right) \quad (11)$$

In Equation (11), the hourly generated power from the solar panel is represented by P_t^{PV}. The area and efficiency of the solar panel are represented by A^{PV} and η^{PV}, respectively. The solar irradiance is represented by Irr, and the hourly temperature of the solar panel is represented by T_t. In our simulations, we considered two solar panels per house, each at 300 W.

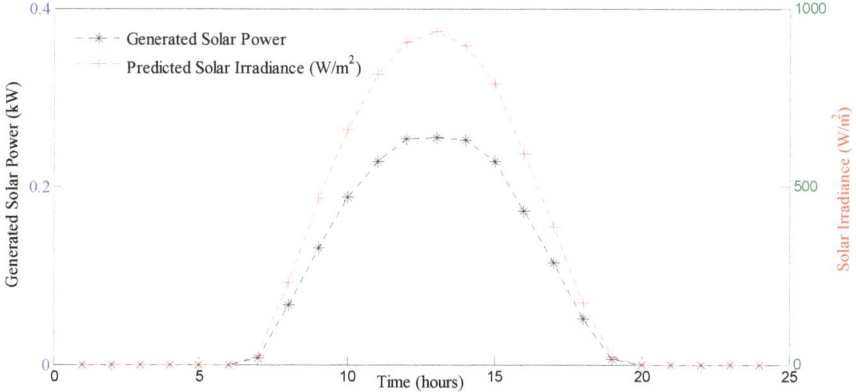

Figure 18. Predicted solar irradiance and power generated by solar panel.

The proportions of the average daily power generated by the wind turbine and solar panels in each month of the year 2007 are presented in Figure 19. It is evident from the figure that the power generated by the solar panels varies predictably across the seasons, as expected. The month of June has the highest recorded average daily power generation using solar panels. The average daily power generated by the wind turbine is lowest in August and saw its best month in February.

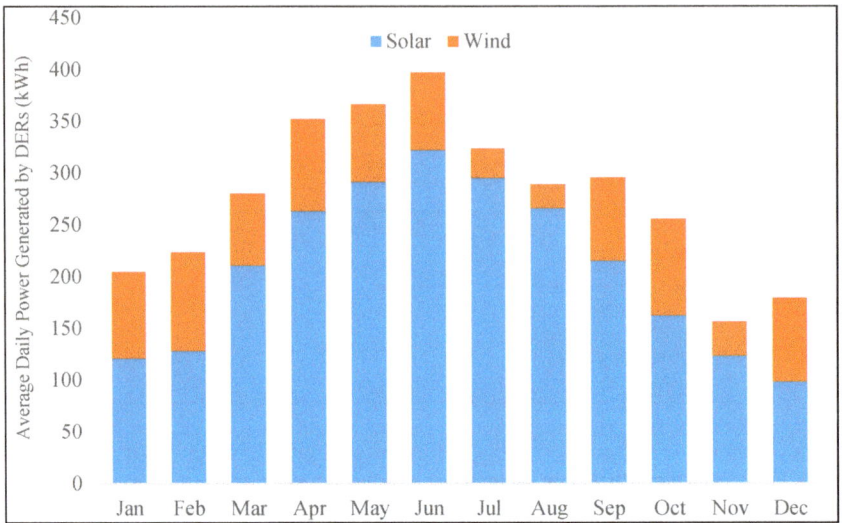

Figure 19. Average daily power generated by solar and wind methods.

The net power load, power generated by DERs, and electricity purchased from the utility are shown in Figure 20. We took the average monthly electricity usage (net load) of households in San Francisco, California, from a previous study [73]. In this figure, we can observe that the total power generated by DERs is high during the months of May and June. This resulted in lower electricity costs when purchased from the utility. Various components of the average monthly cost of electricity for 80 homes are shown in Figure 21. As shown in the figure, savings are at a maximum in the month of June and at a minimum in the month of January.

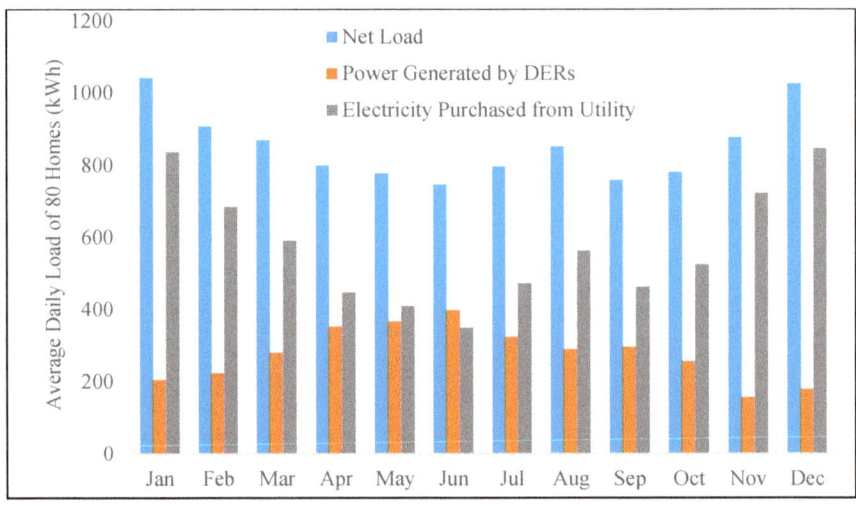

Figure 20. Average daily power generated by DERs and net load.

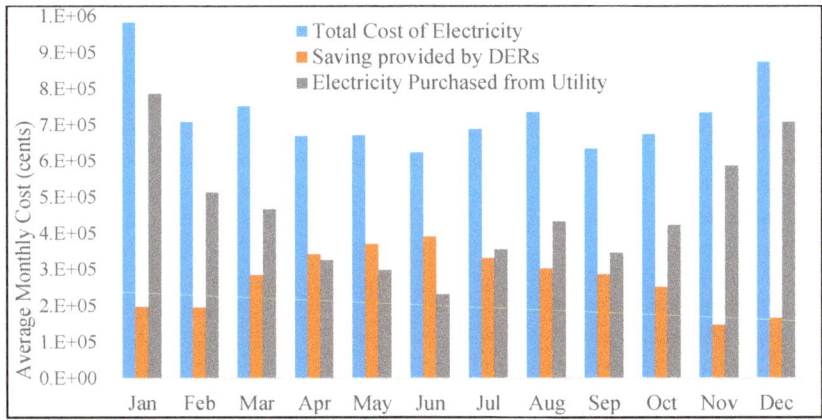

Figure 21. Average monthly cost of electricity for 80 homes and savings provided by DERs.

It can be observed from Figure 22 that the contribution of DERs during the winter season is not very significant. However, in the remaining seasons, significant relief is provided by the DERs in terms of bill reduction. Based on the numerical calculations of the bar graphs in Figure 22, it can be concluded that the proposed framework will help the smart community to achieve an annual reduction of up to 38% in their electricity bills by installing DERs.

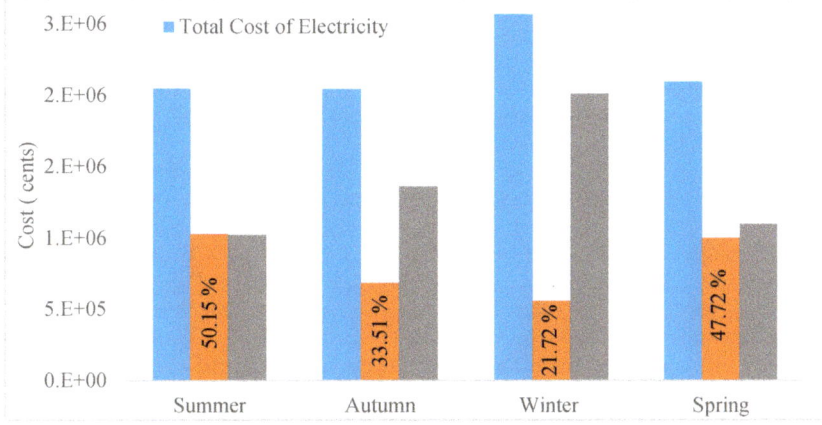

Figure 22. Seasonal cost savings provided by DERs.

6. Conclusions

DERs are valuable in decreasing consumers' electricity bills by enabling them to generate their own green energy. However, the intermittent nature of DERs is a significant issue in accurately forecasting the amount of generated energy through these renewable energy resources. In this work, we proposed and evaluated an efficient approach to energy management in a smart community based on the integration of DERs. Sometimes, the energy generated through DERs is greater than the energy demand of the consumers, which results in a demand and generation mismatch.

The demand and generation mismatch leads to the well-known "duck curve" problem. We applied a machine learning model to accurately predict the generated energy through DERs. We considered a smart community consisting of 80 smart homes. The smart community has access to electric power through a microgrid that is equipped with DERs in the form of wind turbines and photovoltaic systems, in addition to having access to power from the main grid.

The simulation results indicated that our proposed framework is appropriate for approximating the energy generated through DERs and for reducing the electricity bills of the smart community. We evaluated the performance of several machine learning models for selecting a baseline model. Then, we evaluated the performance of our proposed model and compared it with the baseline model.

For the case of wind speed prediction, we obtained 44.94%, 46.12%, and 2.25% error reductions in the evaluation metrics of RMSE, MAE, and sMAPE, respectively. In the case of solar irradiance prediction, we obtained 7.6%, 54.3%, and 0.14% error reductions in the evaluation metrics of RMSE, MBE, and sMAPE, respectively.

We further evaluated the effectiveness of the proposed model in different climates and for different time horizons. The results conclude that the proposed model is not only suitable for short-term forecasting of wind speed and solar irradiance for different time horizons (up to four hours) but for different climates as well.

Author Contributions: Conceptualization, M.A., S.I.H. and K.A.; Methodology, M.A., S.I.H. and K.A.; Validation, S.I.H. and K.A.; Formal analysis, S.I.H. and K.A.; Investigation, M.A., S.I.H. and K.A.; Data curation, S.I.H. and K.A; Writing—original draft, K.A., M.A., and S.I.H.; Writing—review & editing, M.A., S.I.H. and K.A.; Project administration, M.A.; Funding acquisition, K.A.

Funding: The authors extend their appreciation to the Deanship of Scientific Research at King Saud University for funding this work through research group NO (RG-1438-034).

Conflicts of Interest: The authors declare no conflicts of interest.

References

1. Commission Final Report: Integrated Energy Policy Report 2004 Update Publication #100-04-006CM. Available online: https://www.energy.ca.gov/2004_policy_update/ (accessed on 14 April 2019).
2. Engeland, K.; Borga, M.; Creutin, J.-D.; François, B.; Ramos, M.-H.; Vidal, J.-P. Space-time variability of climate variables and intermittent renewable electricity production—A review. *Renew. Sustain. Energy Rev.* **2017**, *79*, 600–617. [CrossRef]
3. Martin, L.; Zarzalejo, L.; Polo, J.; Navarro, A.; Marchante, R.; Cony, M. Prediction of global solar irradiance based on time series analysis: Application to solar thermal power plants energy production planning. *Sol. Energy* **2010**, *84*, 1772–1781. [CrossRef]
4. Falayi, E.; Adepitan, J.; Rabiu, A. Empirical models for the correlation of global solar radiation with meteorological data for Iseyin, Nigeria. *Phys. Sci.* **2008**, *3*, 210–216.
5. Paolik, C.; Voyant, C.; Muselli, M.; Nivet, M. Solar radiation forecasting using ad-hoc time series preprocessing and neural networks. In Proceedings of the 5th International Conference on Emerging Intelligent Computing Technology and Applications, Ulsan, Korea, 16–19 September 2009; pp. 898–907.
6. Paulescu, M.; Paulescu, E.; Gravila, P.; Badescu, V. Modeling solar radiation at the earth surface. In *Weather Modeling and Forecasting of PV Systems Operation*; Springer: Berlin, Germany, 2012.
7. Liu, L.; Liu, D.; Sun, Q.; Li, H.; Wennersten, R. Forecasting power output of photovoltaic system using a BP network method. *Energy Procedia* **2017**, *142*, 780–786. [CrossRef]
8. Bouktif, S.; Fiaz, A.; Ouni, A.; Serhani, M.A. Optimal deep learning LSTM model for electric load forecasting using feature selection and genetic algorithm: Comparison with machine learning approaches. *Energies* **2018**, *11*, 1636. [CrossRef]
9. Hossain, M.; Mekhilef, S.; Danesh, M.; Olatomiwa, L.; Shamshirband, S. Application of extreme learning machine for short term output power forecasting of three grid-connected PV systems. *J. Clean. Prod.* **2017**, *167*, 395–405. [CrossRef]
10. Bendu, H.; Deepak, B.B.V.L.; Murugan, S. Multi-objective optimization of ethanol fuelled HCCI engine performance using hybrid GRNN–PSO. *Appl. Energy* **2017**, *187*, 601–611. [CrossRef]
11. Deo, R.C.; Wen, X.; QiA, F. Wavelet-coupled support vector machine model for forecasting global incident solar radiation using limited meteorological dataset. *Appl. Energy* **2016**, *168*, 568–593. [CrossRef]

12. Lin, S.; Liu, X.; Fang, J.; Xu, Z. Is extreme learning machine feasible? A theoretical assessment (Part II). *IEEE Trans. Neural Netw. Learn. Syst.* **2015**, *26*, 21–34. [CrossRef] [PubMed]
13. Mareček, J. Usage of Generalized Regression Neural Networks in Determination of the Enterprise's Future Sales Plan. *Littera Scr.* **2016**, *3*, 32–41.
14. Bhavsar, H.; Panchal, M.H. A Review on Support Vector Machine for Data Classification. *Int. J. Adv. Res. Comput. Eng. Technol.* **2012**, *1*, 185–189.
15. Shireen, T.; Shao, C.; Wang, H.; Li, J.; Zhang, X.; Li, M. Iterative multi-task learning for time-series modeling of solar panel PV outputs. *Appl. Energy* **2018**, *21*, 654–662. [CrossRef]
16. Wang, F.; Zhen, Z.; Mi, Z.; Sun, H.; Su, S.; Yang, G. Solar irradiance feature extraction and support vector machines based weather status pattern recognition model for short-term photovoltaic power forecasting. *Energy Build.* **2015**, *86*, 427–438. [CrossRef]
17. Ramsami, P.; Oree, V. A hybrid method for forecasting the energy output of photovoltaic systems. *Energy Convers. Manag.* **2015**, *95*, 406–413. [CrossRef]
18. Xiea, J.; Lib, H.; Maa, Z.; Sunc, Q.; Wallinb, F.; Sid, Z.; Gu, J. Analysis of key factors in heat demand prediction with neural networks. *Energy Procedia* **2017**, *105*, 2965–2970. [CrossRef]
19. Ma, Z.; Hue, J.H.; Leijon, A.; Tan, Z.H.; Yang, Z.; Guo, J. Decorrelation of neutral vector variables: Theory and applications. *IEEE Trans. Neural Netw.* **2016**, *29*, 129–143. [CrossRef]
20. Cococcioni, M.; D'Andrea, E.; Lazzerini, B. Twenty-four-hour-ahead forecasting of energy production in solar PV systems. In Proceedings of the IEEE International Conference on Intelligent Systems Design and Applications, Cordoba, Spain, 22–24 November 2011; pp. 1276–1281.
21. Chu, Y.; Urquhart, B.; Gohari, S.M.I.; Pedro, H.T.C.; Kleissl, J.; Coimbra, C.F.M. Short-term reforecasting of power output from a 48 MW solar PV plant. *Sol. Energy* **2015**, *112*, 68–77. [CrossRef]
22. Hussain, S.; AlAlili, A. Hybrid solar radiation modelling approach using wavelet multiresolution analysis and artificial neural networks. *Appl. Energy* **2017**, *208*, 540–550. [CrossRef]
23. Larraondo, P.R.; Inza, I.; Lozano, J.A. Automating weather forecasts based on convolutional networks. In Proceedings of the ICML Workshop on Deep Structured Prediction, PMLR 70, Sydney, Australia, 11 August 2017.
24. Zhuang, W.Y.; Ding, W. Long-lead prediction of extreme precipitation cluster via a spatio-temporal convolutional neural network. In Proceedings of the 6th International Workshop on Climate Informatics: CI 2016, Boulder, CO, USA, 22–23 September 2016.
25. Sulagna Gope, P.M.; Sarkar, S. Prediction of extreme rainfall using hybrid convolutional-long short term memory networks. In Proceedings of the 6th International Workshop on Climate Informatics, Boulder, CO, USA, 22–23 September 2016.
26. Han, Q.; Wu, H.; Hu, T.; Chu, F. Short-term wind speed forecasting based on signal decomposing algorithm and hybrid linear/nonlinear models. *Energies* **2018**, *11*, 2976. [CrossRef]
27. Rabanal, A.; Ulazia, A.; Ibarra-Berastegi, G.; Sáenz, J.; Elosegui, U. MIDAS: A Benchmarking Multi-Criteria Method for the Identification of Defective Anemometers in Wind Farms. *Energies* **2019**, *12*, 28. [CrossRef]
28. Huang, C.-J.; Kuo, P.-H. A Short-Term Wind Speed Forecasting Model by Using Artificial Neural Networks with Stochastic Optimization for Renewable Energy Systems. *Energies* **2018**, *11*, 2777. [CrossRef]
29. Ssekulima, E.B.; Anwar, M.B.; Al Hinai, A.; El Moursi, M.S. Wind speed and solar irradiance forecasting techniques for enhanced renewable energy integration with the grid: A review. *IET Renew. Power Gener.* **2016**, *7*, 885–989. [CrossRef]
30. Yahyaoui, I. *Forecasting of Intermittent Solar Energy Resource, Advances in Renewable Energies and Power Technologies*, 1st ed.; Volume 1: Solar and Wind Energies; Elsevier Science: Amsterdam, The Netherlands, 2018; Chapter 3; pp. 78–114.
31. Coimbra Carlos, F.M.; Kleissl, J.; Marquez, R. Overview of Solar-Forecasting Methods and a Metric for Accuracy Evaluation. In *Solar Energy Forecasting and Resource Assessment*; Elsevier Academic Press: Amsterdam, The Netherlands, 2013; Chapter 8, pp. 171–194.
32. Zhang, J.; Florita, A.; Hodge, B.-M.; Lu, S.; Hamann, H.F.; Banunarayanan, V.; Brockway, A.M. A suite of metrics for assessing the performance of solar power forecasting. *Sol. Energy* **2015**, *111*, 157–175. [CrossRef]
33. Tuohy, A.; Zack, J.; Haupt, S.E.; Sharp, J.; Ahlstrom, M.; Dise, S.; Grimit, E.; Mohrlen, C.; Lange, M.; Casado, M.G.; et al. Solar Forecasting: Methods, Challenges, and Performance. *IEEE Power Energy Mag.* **2015**, *13*, 50–59. [CrossRef]

34. Reikard, G.; Hansen, C. Forecasting solar irradiance at short horizons: Frequency and time domain models. *Renew. Energy* **2019**, *135*, 1270–1290. [CrossRef]
35. Kleissl, J. *Current State of the Art in Solar Forecasting*; Technical Report; California Institute for Energy and Environment: Berkeley, CA, USA, 2010.
36. Kumler, A.; Xie, Y.; Zhang, Y. *A New Approach for Short-Term Solar Radiation Forecasting Using the Estimation of Cloud Fraction and Cloud Albedo*; NREL/TP-5D00-72290; National Renewable Energy Laboratory (NREL): Golden, CO, USA, 2018.
37. Pedro Hugo, T.C.; Coimbra Carlos, F.M.; David, M.; Lauret, P. Assessment of machine learning techniques for deterministic and probabilistic intra-hour solar forecasts. *Renew. Energy* **2018**, *123*, 191–203. [CrossRef]
38. Ma, J.; Makarov, Y.V.; Loutan, C.; Xie, Z. Impact of wind and solar generation on the California ISO's intra-hour balancing needs. In Proceedings of the 2011 IEEE Power and Energy Society General Meeting, San Diego, CA, USA, 24–29 July 2011.
39. Huang, R.; Huang, T.; Gadh, R. Solar Generation Prediction using the ARMA Model in a Laboratory-level Micro-grid. In Proceedings of the 2012 IEEE Third International Conference on Smart Grid Communications, Tainan, Taiwan, 5–8 November 2012; pp. 528–533.
40. Bouzgou, H.; Gueymard, C.A. Fast short-term global solar irradiance forecasting with wrapper mutual information. *Renew. Energy* **2019**, *133*, 1055–1065. [CrossRef]
41. Benali, L.; Notton, G.; Fouilloy, A.; Voyant, C.; Dizene, R. Solar radiation forecasting using artificial neural network and random forest methods: Application to normal beam, horizontal diffuse and global components. *Renew. Energy* **2019**, *132*, 871–884. [CrossRef]
42. Crisosto, C.; Hofmann, M.; Mubarak, R.; Seckmeyer, G. One-Hour Prediction of the Global Solar Irradiance from All-Sky Images Using Artificial Neural Networks. *Energies* **2018**, *11*, 2906. [CrossRef]
43. Lago, J.; De Brabandere, K.; De Ridder, F.; De Schutter, B. Short-term forecasting of solar irradiance without local telemetry: A generalized model using satellite data. *Sol. Energy* **2018**, *173*, 566–577. [CrossRef]
44. Jadidi, A.; Menezes, R.; De Souza, N.; Lima, A.C.D. A Hybrid GA-MLPNN Model for One-Hour-Ahead Forecasting of the Global Horizontal Irradiance in Elizabeth City, North Carolina. *Energies* **2018**, *11*, 2641. [CrossRef]
45. Alfadda, A.; Rahman, S.; Pipattanasomporn, M. Solar irradiance forecast using aerosols measurements: A data driven approach. *Sol. Energy* **2018**, *170*, 924–939. [CrossRef]
46. California Energy Commission Report: Toward a Clean Energy Future. Publication # CEC-100-2018-001-V1. Available online: https://www.energy.ca.gov/2018publications/CEC-100-2018-001/. (accessed on 14 April 2019).
47. NSRD from NREL. Available online: https://nsrdb.nrel.gov/current-version (accessed on 10 March 2019).
48. Sengupta, M.; Xie, Y.; Lopez, A.; Habte, A.; Maclaurin, G.; Shelby, J. The National Solar Radiation Data Base (NSRDB). *Renew. Sustain. Energy Rev.* **2018**, *89*, 51–60. [CrossRef]
49. Aslam, S.; Iqbal, Z.; Javaid, N.; Khan, Z.A.; Aurangzeb, K.; Haider, S.I. Towards efficient energy management of smart buildings exploiting heuristic optimization with real time and critical peak pricing schemes. *Energies* **2017**, *12*, 2065. [CrossRef]
50. Mohsenian-Rad, A.-H.; Wong, V.W.S.; Jatskevich, J.; Schober, R.; Leon-Garcia, A. Autonomous demand-side management based on game-theoretic energy consumption scheduling for the future smart grid. *IEEE Trans. Smart Grid* **2010**, *1*, 320–331. [CrossRef]
51. Niyato, D.; Xiao, L.; Wang, P. Machine-to-machine communications for home energy management system in smart grid. *IEEE Commun. Mag.* **2011**, *49*, 53–59. [CrossRef]
52. Lidula, N.; Rajapakse, A.D. Microgrids research: A review of experimental microgrids and test systems. *Renew. Sustain. Energy Rev.* **2011**, *15*, 186–202. [CrossRef]
53. Mehrizi-Sani, A.; Iravani, R. Potential-function based control of a microgrid in islanded and grid-connected modes. *IEEE Trans. Power Syst.* **2010**, *25*, 1883–1891. [CrossRef]
54. Farhangi, H. The path of the smart grid. *IEEE Power Energy Mag.* **2010**, *8*, 18–28. [CrossRef]
55. Lasseter, R.H.; Akhil, A.; Marnay, C.; Stephens, J.; Dagle, J.; Guttromson, R.; Meliopoulous, A.; Yinger, R.; Eto, J. The CERTS Microgrid Concept: White Paper on Integration of Distributed Energy Resources. California Energy Commission, Office of Power Technologies, U.S. Department of Energy; LBNL-50829. Available online: http://certs.lbl.gov (accessed on 14 April 2019).

56. Lasseter, R.H.; Paigi, P. Microgrid: A conceptual solution. In Proceedings of the IEEE Power Electronics Specialists Conference (PESC 2004), Aachen, Germany, 20–25 June 2004.
57. Maffei, A.; Srinivasan, S.; Castillejo, P.; Martínez, J.F.; Iannelli, L.; Bjerkan, E.; Glielmo, L. A semantic-middleware-supported receding horizon optimal power flow in energy grids. *IEEE Trans. Ind. Inform.* **2018**, *14*, 35–46. [CrossRef]
58. Athraa, A.K.; Izzri, A.W.N.; Ishak, A.; Jasronita, J.; Ahmed, A. Advanced wind speed prediction model based on a combination of Weibull distribution and an artificial neural network. *Energies* **2017**, *10*, 1744. [CrossRef]
59. Pusat, S.; Akkoyunlu, M.T. Effect of time horizon on wind speed prediction with ANN. *J. Therm. Eng.* **2018**, *4*, 1770–1779. [CrossRef]
60. Liu, J.N.K.; Hu, Y.; He, Y.; Chan, P.W.; Lai, L. Deep neural network modeling for big data weather forecasting. In *Information Granularity, Big Data, and Computational Intelligence: Studies in Big Data*; Springer: Berlin, Germany, 2014; Volume 8, pp. 389–408.
61. Gensler, A.; Henze, J.; Sick, B.; Raabe, N. Deep Learning for solar power forecasting: An approach using auto encoder and LSTM neural networks. In Proceedings of the IEEE International Conference on Systems, Man, and Cybernetics (SMC), Budapest, Hungary, 9–12 October 2016; pp. 2858–2865.
62. Hasan, M.S.; Kouki, Y.; Ledoux, T.; Pazat, J.-L. Exploiting renewable sources: When green SLA becomes a possible reality in cloud computing. *IEEE Trans. Cloud Comput.* **2017**, *5*, 249–262. [CrossRef]
63. Chaudhary, P.; Rizwan, M. Energy management supporting high penetration of solar photovoltaic generation for smart grid using solar forecasts and pumped hydro storage system. *Renew. Energy* **2018**, *118*, 928–946. [CrossRef]
64. Khan, M.; Javaid, N.; Javaid, S.; Aurangzeb, K. Kernel based support vector quantile regression for real-time data analysis. In Proceedings of the International Conference on Innovation and Intelligence for Informatics, Computing, and Technologies, Zallaq, Bahrain, 18–20 November 2018, accepted for publication.
65. Denholm, P.; O'Connell, M.; Brinkman, G.; Jorgenson, J. *Overgeneration from solar energy in California: A Field Guide to the Duck Chart*; NREL/TP-6A20-65023; National Renewable Energy Laboratory (NREL): Golden, CO, USA, 2015; pp. 1–46.
66. Rehman, O.U.; Khan, S.A.; Malik, M.; Javaid, N.; Javaid, S.; Aurangzeb, K. Optimal scheduling of distributed energy resources for load balancing and user comfort management in smart grid. In Proceedings of the International Conference on Innovation and Intelligence for Informatics, Computing, and Technologies, Zallaq, Bahrain, 18–20 November 2018. accepted for publication.
67. Weather Data. Available online: https://maps.nrel.gov/nsrdb-viewer/ (accessed on 31 December 2018).
68. Time of Use Tariff of California. Available online: https://www.sce.com/residential/rates/Time-Of-Use-Residential-Rate-Plans (accessed on 2 December 2018).
69. Khalid, A.; Aslam, S.; Aurangzeb, K.; Haider, S.I.; Ashraf, M.; Javaid, N. An efficient energy management approach using fog-as-a-service for sharing economy in a smart grid. *Energies* **2018**, *11*, 3500. [CrossRef]
70. Details of Selected Wind Turbine. Available online: https://en.wind-turbine-models.com/turbines/1864-aeolos-aeolos-h-30kw (accessed on 28 February 2019).
71. Billinton, R.; Bai, G. Generating capacity adequacy associated with wind energy. *IEEE Trans. Energy Convers.* **2004**, *1*, 641–646. [CrossRef]
72. Solar Panel Temperature Data. Available online: https://pvwatts.nrel.gov/pvwatts.php (accessed on 2 December 2018).
73. Average Monthly Electricity Usage of San Francisco, California, Household. Available online: https://energycenter.org/equinox/dashboard/residential-electricity-consumption (accessed on 13 January 2019).

© 2019 by the authors. Licensee MDPI, Basel, Switzerland. This article is an open access article distributed under the terms and conditions of the Creative Commons Attribution (CC BY) license (http://creativecommons.org/licenses/by/4.0/).

Article

Combining Weather Stations for Electric Load Forecasting

Masoud Sobhani, Allison Campbell, Saurabh Sangamwar, Changlin Li and Tao Hong *

Department of Systems Engineering and Engineering Management, University of North Carolina at Charlotte, 28223 Charlotte, NC, USA; msobhani@uncc.edu (M.S.); acampb79@uncc.edu (A.C.); ssangamw@uncc.edu (S.S.); cli33@uncc.edu (C.L.)
* Correspondence: Tao.Hong@uncc.edu

Received: 18 March 2019; Accepted: 12 April 2019; Published: 21 April 2019

Abstract: Weather is a key factor affecting electricity demand. Many load forecasting models rely on weather variables. Weather stations provide point measurements of weather conditions in a service area. Since the load is spread geographically, a single weather station may not sufficiently explain the variations of the load over a vast area. Therefore, a proper combination of multiple weather stations plays a vital role in load forecasting. This paper answers the question: given a number of weather stations, how should they be combined for load forecasting? Simple averaging has been a commonly used and effective method in the literature. In this paper, we compared the performance of seven alternative methods with simple averaging as the benchmark using the data of the Global Energy Forecasting Competition 2012. The results demonstrate that some of the methods outperform the benchmark in combining weather stations. In addition, averaging the forecasts from these methods outperforms most individual methods.

Keywords: weather station combination; electric load forecasting; hierarchical load forecasting

1. Introduction

Electric load forecasting is an essential input for the decision-making processes in the power industry. In past decades, numerous forecasting models that include calendar and weather variables have been developed and tested [1–3]. A recent review of the load forecasting techniques is presented in [4]. Many power system applications require customized load forecasting efforts. The majority of the papers in load forecasting literature have studied the methodologies of point load forecasting at high voltage levels [5,6]. Deployment of the smart grid technologies in the recent years, along with the dispersion of renewable energy and electric vehicles, has necessitated new solutions such as probabilistic load forecasting and created the need for accurate forecasting at low/medium voltage levels [7–9].

Weather is a key driving factor in electricity consumption. Weather-based models have been used frequently for electric load forecasting. Forecasters have employed the correlation between weather and load profiles to develop models. Although temperature is a frequently used weather variable, others such as relative humidity and wind speed have been used in load forecasting models as well [10,11]. Weather data mainly come from the observations at weather stations. While many public data providers and private vendors obtain data from different weather stations to serve their customers in the power industry, the availability and quality of weather data has been a concern for power companies [12].

The instruments of a weather station typically collect the information from a limited geographic area. The data such as temperature or humidity reflect the weather behavior of a specific location. On the other hand, load is distributed in a vast area of a service zone. In a power grid, as we move to the higher levels of load hierarchy, the aggregated load covers a larger geographic region. Therefore, the point readings of a single weather station may not sufficiently explain the load variations over a vast area.

Typically, multiple weather stations are inside or around the service territory, which leaves the load forecasters with the question of how to best utilize the weather data collected from different stations. Although a forecasting model may use multiple temperature profiles simultaneously [5], most load forecasting models in the literature use a single weather profile to predict the load. Therefore, selecting and combining the temperature profiles from a group of stations is crucial to the performance of the forecasting models. If only one weather station is selected amongst several, that would limit the opportunity to use all available data [13]. A simple average of the weather stations has been used frequently in the literature. Hong et al. proposed a comprehensive weather station selection and combination methodology, where the weather stations are ranked based on in-sample fit error and then combined by taking a simple average [14]. In [15] each weather station was used to generate a unique load forecast and then the exponentially weighted average algorithm combines the forecasts and choses the best combination based on the forecast accuracy. A similar approach was used in [16], where singular value decomposition was used to weight each forecast in the final combination. In addition, some vendors such as Meteo French provide national average weather profiles by weighting the data from different regions [17].

Despite different methods to combine weather stations, the load forecasting literature has not yet offered a formal comprehensive comparison. Lai and Hong [18] showed that temperatures weighted by economy and load are not necessarily better than the equal-weight combination, a.k.a. taking a simple average of temperatures from all stations. In this paper, we used the equal-weight combination as the benchmark. We compared the performance of seven combination methods with the benchmark. Four of the seven methods were based on simple concepts, such as linear weights, exponential weights, performance-based weights derived from mean absolute percentage error (MAPE), and geometric mean. We also propose two other methods including a twofold combination method and a genetic algorithm (GA) based method. The seventh method is an ensemble by averaging these six individual forecasts and the benchmark. These seven methods together with the benchmark were evaluated through an empirical study constructed using the data of the Global Energy Forecasting Competition 2012 (GEFCom2012) [19]. While most methods outperformed the equal-weight combination, the ensemble appears to be the most robust and accurate one on average.

The paper is organized as follows: Section 2 introduces the seven methods; Section 3 presents the data, experiments, and the discussion; the paper is concluded in Section 4.

2. Methodology

The creation of a single temperature time series from multiple weather stations, a virtual weather station [14], is crucial to load forecast accuracy. It involves two components: weather station selection and weather station combination. Ideally, these two components should be executed simultaneously. Implementing only one of them or executing both components sequentially is likely to lead to a suboptimal result. In this paper, we focus on the combination component only to avoid distractions with fine-tuning.

This work builds on the weather station selection methodology proposed in [14] by focusing on different combination methods. Therefore, we used the same methodology proposed in [14] to select the weather stations. In [14], the weather stations were ranked based on the in-sample fit error and sorted accordingly in ascending order. In the next step, the top ranked weather stations were combined into virtual weather stations via simple averaging. The threshold for the number of top weather stations was selected based on an out-of-sample test in the validation year. In this paper, we assumed that the weather stations were selected by this algorithm and we only addressed weather station combination. The complete proposed methodology of [14] is the benchmark of our experiments.

There are many ways to create virtual weather stations. Taking a simple average of the weather stations, or weighting the stations equally, is a practical and straightforward approach that has been used frequently in the past [14,18]. In this paper, we tested the efficacy of seven alternatives to the equal-weight combination. The seven alternatives can be broken down to three categories: simple methods, complex weighting methods, and an ensemble.

The proposed combination methods were tested for their application to electricity load forecasting. Tao's Vanilla Benchmark model is a frequently cited load forecasting model that has been used in many forecasting competitions and studies [14,19,20]. This model was used to create load forecasts in this work. The Vanilla Benchmark model is a weather-based load forecasting model that employs polynomials of the temperature and their interactions with calendar variables to predict load. The model can be specified as follows:

$$L_t = \beta_0 + \beta_1 Trend + \beta_2 M_t + \beta_3 W_t + \beta_4 H_t + \beta_5 W_t H_t + \beta_6 T_t + \beta_7 T_t^2 + \beta_8 T_t^3 + \beta_9 M_t T_t \\ + \beta_{10} M_t T_t^2 + \beta_{11} M_t T_t^3 + \beta_{12} H_t T_t + \beta_{13} H_t T_t^2 + \beta_{14} H_t T_t^3 \tag{1}$$

where L_t is the load forecast for time t; β_i are the coefficients estimated using the ordinary least square method; M_t, W_t and H_t are the coincident month-of-the-year, day-of-the-week, and hour-of-the-day for time t, respectively, which are classification variables; and T_t is the coincident temperature.

We used Mean Absolute Percentage Error (MAPE) for evaluation. MAPE is expressed as follows:

$$\text{MAPE} = \frac{100}{n} \sum_{t=1}^{n} \left| \frac{L_t - \hat{L}_t}{L_t} \right| \tag{2}$$

where L_t is the actual load; $\hat{L}_t$ is the predicted load; and n is the number of observations.

2.1. Linear Combination

The linear combination method allocates decreasing linear weights to the weather stations ranked in the increasing order of their MAPE values. For example, if we have four weather stations ranked by their corresponding MAPE values in the increasing order, the linear weights assigned to these stations are 4, 3, 2, and 1, respectively. We then normalized the weights so that the sum of the weights equal to one in order to keep the combined temperature profile in the same range as the individual ones. Let lin_w_i be the linear weight, w_i be the normalized weight, and n be the total number of weather stations. The normalized weights are calculated as follows:

$$w_i = \frac{lin\_w_i}{\sum_{i=1}^{n} lin\_w_i} \tag{3}$$

where

$$lin\_w_i = \begin{cases} n & i = 1 \\ n - 1 & i = 2 \\ \vdots & \vdots \\ 1 & i = n \end{cases} \tag{4}$$

2.2. Exponential Combination

The exponential combination method assigns the exponentially decaying weights to the weather stations ranked by their MAPE values from small to large. Let exp_w_i be the exponential weight, and b be the base. The normalized weight is expressed as follows:

$$w_i = \frac{exp\_w_i}{\sum_{i=1}^{n} exp\_w_i} \tag{5}$$

where

$$exp\_w_i = \begin{cases} b^n & i=1 \\ b^{n-1} & i=2 \\ \vdots & \vdots \\ b^1 & i=n \end{cases} \quad (6)$$

2.3. MAPE-Based Combination

The MAPE-based combination method uses the MAPE value of a weather station as its weight. Let $mape\_w_i$ be the MAPE value of weather station i. The normalized weight is expressed as follows:

$$w_i = \frac{100 - mape\_w_i}{\sum_{i=1}^{n}(100 - mape\_w_i)} \quad (7)$$

2.4. Geometric Mean Combination

The geometric mean of n numbers is the nth root of their product. It indicates the central tendency of a set of numbers. The geometric mean combination method calculates the geometric mean of the temperature series of n weather stations as follows:

$$T_{gmean} = \sqrt[n]{T_1 T_2 \ldots T_n} \quad (8)$$

where T_i is the temperature profile of the weather station i.

2.5. Twofold Combination

The methodology proposed in [14] creates each virtual station by taking a simple average of the top ranked stations. The twofold combination method takes one more iteration to generate secondary virtual stations by combining the virtual stations created in [14]. By doing twofold combination, we magnified the role of top ranked stations in the second blend. The method can be implemented as follows:

(1) Rank the original stations based on the ascending order of their in-sample fit error of the load forecasting model.
(2) Create virtual stations by taking the simple averages of top stations.
(3) Forecast the validation year using each virtual temperature profile, and calculate MAPE for each forecast.
(4) Sort the virtual stations based on the MAPE of the validation year in ascending order.
(5) Create the secondary virtual stations by taking the simple averages of top virtual stations.
(6) Forecast the validation year again using the temperature profile of each secondary virtual station, and calculate MAPE for each forecast.
(7) The secondary virtual station corresponding to the smallest MAPE value provides the desired temperature profile.

Figure 1 shows the process of twofold combination in a flowchart.

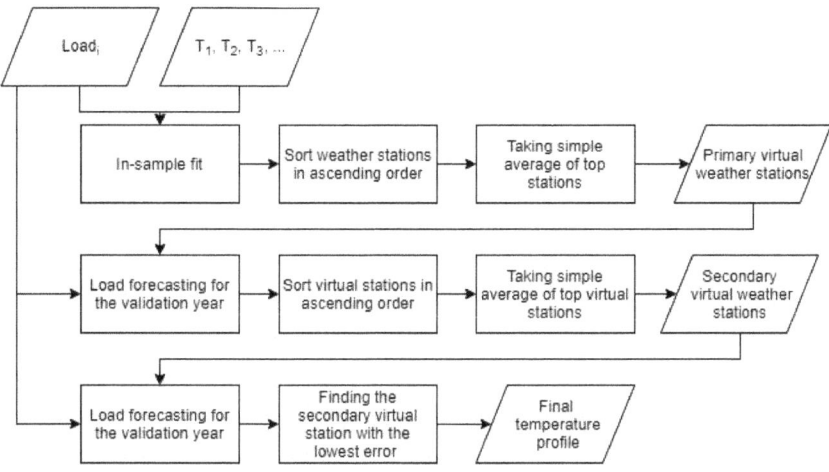

Figure 1. Twofold combination flowchart.

2.6. GA-Based Combination

Inspired by the natural selection in biological evolution, genetic algorithms are well-suited to solve an optimization problem. Considering weather station combination as an optimization problem, we can apply GA to find the weights that can minimize forecast errors. The methodology can be broken down into five steps:

(1) Initialize a population of individuals as a string of randomly assigned weights, where each is assigned a weight, and the population of individuals captures a spectrum of possible weights for each station.
(2) Create virtual stations using these sets of weights.
(3) Evaluate the goodness of fit for each virtual station using MAPE.
(4) Generate a subsequent population in the evolution, allow individuals in the population to cross and mutate.
(5) After all the designated generations have completed, the desired virtual station will correspond to the weights that lead to the best goodness of fit.

The weighting parameters were generated randomly and optimized with a genetic algorithm based on the OneMax algorithm [21]. Each weather station for a given zone was assigned a position in a list [0, 1, 2, ... , n], and was given a random number between 0 and 1. A population of individuals was initialized. In practice, populations greater than several hundred are recommended. After initialization, the fitness of each individual was calculated, and individuals were "mated" with crossing and mutation probabilities for a given number of generations. In this paper, 500 individuals were initialized, 15 generations were implemented, and the probability of crossing and mutation was set to 50% and 20%, respectively.

The virtual station for each individual was then fed into the load forecasting model. The goodness of fit was evaluated with the MAPE of the validation year. Once each individual had a MAPE value, the genetic algorithm ranked the population, crossed and mutated ("created offspring" for the following generation), and proceeded to the next generation of individuals. The process continued until either of the following conditions were met: (1) the MAPE is below a cutoff threshold, (2) a predefined number of generations have completed. While alternative hyper-parameters may lead to better results, we did not fine-tune them in this paper.

2.7. An Ensemble

The six aforementioned combination methods together with the benchmark method weight the temperature profiles from selected weather stations differently. While they are all heuristics, it is difficult to tell up front which method(s), if any, can dominate the others. Forecast combination has been a best practice in forecasting. Therefore, we can create an ensemble by taking a simple average of all forecasts, with the hope that this ensemble can be more accurate and robust than most individual forecasts.

3. Experiments

3.1. Data

The data used in the experiments are from the load forecasting track of Global Energy Forecasting Competition 2012 [19]. The data consist of the hourly load and hourly temperature of a U.S. utility. The load data include 20 different load zones, while the temperature data come from 11 weather stations across the service territory. Zone 21 is the aggregate load of all 20 load zones. In our case study, we used four years of data. Two years of 2004 and 2005 were used for training, while years 2006 and 2007 were the validation and test years, respectively.

The load data were for the residential and commercial sectors, which typically have a strong correlation with weather variables. Figure 2 shows the scatter plot of temperature vs. load using three years of data (2004 to 2006) from load zone 1 and weather station 6. The graph clearly depicts a strong correlation between the temperature and load. Figure 3 shows the grouped boxplots of load and temperature during the same three years, which illustrate the seasonal patterns of both profiles.

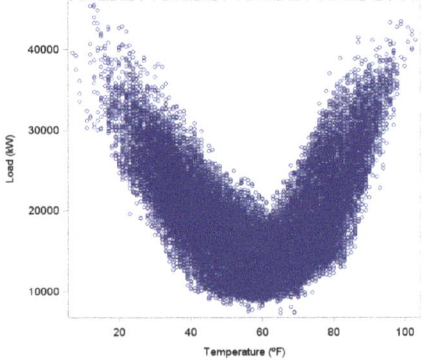

Figure 2. Load-temperature scatter plot.

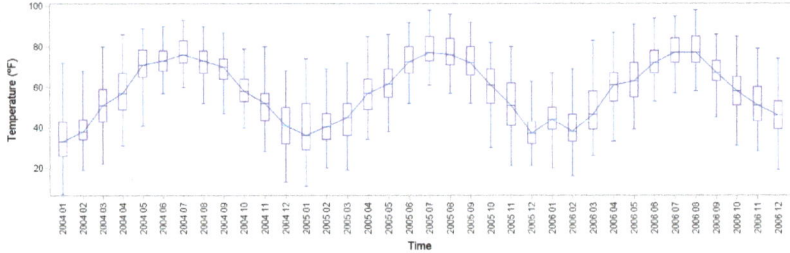

Figure 3. *Cont.*

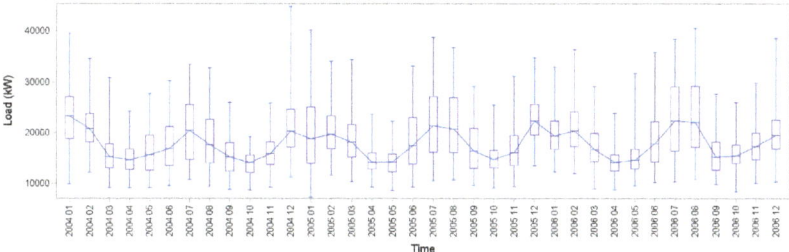

Figure 3. Boxplots of data grouped by month, for temperature (**up**) from station 6 and load (**down**) from zone 1.

Table 1 lists the stations selected for each load zone using the weather station selection methodology proposed by [14]. The weather stations are listed from left to right based on the in-sample fit error in ascending order. In this paper, we propose and evaluate alternative methods to combine these selected weather stations for each load zone.

Table 1. Constrained numbered weather stations for each Global Energy Forecasting Competition 2012 (GEFCom2012) load zone.

Zone	Stations
Z_1	(6, 10, 2)
Z_2	(9, 7, 11, 2, 10, 5)
Z_3	(9, 7, 11, 2, 10, 5)
Z_4	(7, 2)
Z_5	(9, 7, 10)
Z_6	(9, 7, 11, 2, 10, 3, 5)
Z_7	(9, 7, 11, 2, 10, 5)
Z_8	(11, 2, 9, 7)
Z_9	(4, 7, 8, 1, 10, 6, 3, 9, 5)
Z_{10}	(1, 3, 5, 4, 7, 6)
Z_{11}	(3, 5, 7, 1)
Z_{12}	(5, 3, 7, 1)
Z_{13}	(2, 11, 6, 10)
Z_{14}	(4, 6, 8, 10, 1)
Z_{15}	(6, 10)
Z_{16}	(7, 3, 9, 10, 5, 6, 1)
Z_{17}	(4, 8, 6, 10, 7, 1)
Z_{18}	(7)
Z_{19}	(10, 6, 7)
Z_{20}	(11, 2, 9, 7, 10, 6)
Z_{21}	(7, 9, 10, 6, 3, 2, 5, 11, 8, 4, 1)

3.2. Results

The heat map in Table 2 depicts the MAPEs of 21 load zones under all eight different methods including the benchmark from [14] and the seven methods proposed in this paper. A cooler color (green) indicates a lower MAPE value, while a warmer color (red) indicates a higher MAPE value.

Following [14] we exclude Zone 4, which experienced a major outage, and Zone 9, which is an industrial customer not responsive to the weather conditions.

The results show a diverse performance of the weather station combination methods in each load zone. Overall, not a single method dominates all zones. At the aggregate level Zone 21, the GA-based combination performed the best, while only one of the seven alternatives, exponential combination underperformed the benchmark. On average for the 18 regular load zones, only two of the seven alternatives, MAPE-based combination and GA-based combination, underperformed the benchmark. These observations suggest that the benchmark, e.g., weighting the selected weather stations equally, has some more accurate alternatives. The ensemble performed the best on average for the 18 regular zones, improving the benchmark by 0.6% on average. The largest improvement was in zone 10, which was a 3% enhancement. Among the five methods that outperformed the benchmark of the regular zones, the ensemble outperformed the other four in the aggregate zone. Between the benchmark and the ensemble on the 18 regular zones plus the aggregate zone, the ensemble won 15 zones, tied 1 zone, and only lost 3 zones. Considering robustness and accuracy together, the ensemble is the best alternative, though it also involves more computational efforts than its counterparts.

Table 2. Mean absolute percentage errors (MAPEs) of the load zones under different weather station combination methods.

	Zone	No.	Benchmark	Linear	Exponential	MAPE	G-mean	Twofold	GA	Ensemble
Aggregate	21	11	5.22	5.16	5.28	5.13	5.19	5.22	5.06	5.14
Regular Zones	1	3	7.01	7.17	7.21	7.00	6.99	7.01	7.03	7.04
	2	6	5.62	5.65	5.70	5.62	5.60	5.62	5.59	5.61
	3	6	5.62	5.65	5.70	5.62	5.60	5.62	5.59	5.61
	5	3	9.88	9.56	9.45	9.92	9.85	9.88	10.61	9.84
	6	7	5.55	5.57	5.62	5.56	5.54	5.54	5.54	5.53
	7	6	5.62	5.65	5.70	5.62	5.60	5.62	5.59	5.61
	8	4	7.50	7.41	7.38	7.52	7.48	7.50	7.37	7.43
	10	6	6.70	6.45	6.36	6.72	6.68	6.39	6.46	6.49
	11	4	7.70	7.67	7.69	7.72	7.69	7.70	7.68	7.67
	12	4	6.78	6.62	6.57	6.84	6.76	6.78	6.80	6.69
	13	4	7.39	7.27	7.24	7.40	7.36	7.39	7.51	7.35
	14	5	9.38	9.34	9.27	9.40	9.39	9.38	9.33	9.31
	15	2	7.44	7.42	7.42	7.44	7.44	7.44	7.43	7.43
	16	7	8.12	8.14	8.13	8.14	8.10	8.08	8.15	8.04
	17	6	5.26	5.37	5.52	5.26	5.27	5.26	5.28	5.29
	18	1	6.72	6.72	6.72	6.72	6.72	6.72	6.72	6.72
	19	3	7.90	7.97	7.98	7.90	7.90	7.90	7.92	7.91
	20	6	5.75	5.63	5.61	5.76	5.73	5.72	5.61	5.66
2–11 Average			7.00	6.96	6.96	7.01	6.98	6.97	7.01	6.96
Excluded Zones	4	2	16.08	16.27	16.27	16.08	16.05	16.08	16.12	16.13
	9	9	139.16	139.95	141.73	139.16	139.15	139.17	138.56	139.71

4. Conclusions

Weather variables have been used in many load forecasting models. The electric load profiles of residential and commercial customers have shown significant correlation between weather and load. Power companies use weather data as a major source of information to predict future demand. Weather properties such as temperature and relative humidity are provided by the weather stations located in a service zone. Because some service territories are large, some utilities rely on multiple weather stations. Combining multiple weather profiles helps the forecasting model explain more variations of the load spread.

This paper compared several combination methods to address this challenge. Seven different combination methods were compared with the benchmark simple averaging, which was used in [14], through a case study using the data from the load forecasting track of GEFCom2012. The results suggested that several alternatives are more accurate than combining the stations with equal weights. This could be due to many factors such as the distribution of the load data, the geographic distribution of the service territory and the location and number of weather stations. In addition, the ensemble that takes a simple average of different combination methods offers the most robust performance and accurate forecasts.

There are numerous papers that discuss the importance of weather data for load forecasting. This paper is among the first to formally investigate and evaluate different methods for weather station combination. The proposed combination methods are some representative techniques. These techniques could be improved and customized to enhance the load forecasting performance. Future research directions may include other factors such as location information into a combination method to capture the benefits of having multiple weather stations.

Author Contributions: T.H. supervised the project and designed the experiment. M.S., A.C., S.S., and C.L. designed the various methods. M.S. coordinated the reproduction of all methods. M.S. drafted the original version. A.C. and T.H. revised the paper.

Funding: This work was supported in part by the US Department of Energy, Cybersecurity for Energy Delivery Systems (CEDS) Program under Work Order Number M616000124.

Conflicts of Interest: The authors declare no conflict of interest.

References

1. Hippert, H.S.; Pedreira, C.E.; Souza, R.C. Neural networks for short-term load forecasting: A review and evaluation. *IEEE Trans. Power Syst.* **2001**, *16*, 44–55. [CrossRef]
2. Xie, J.; Hong, T. Load forecasting using 24 solar terms. *J. Mod. Power Syst. Clean Energy* **2018**, *6*, 208–214. [CrossRef]
3. Xie, J.; Hong, T. Temperature Scenario Generation for Probabilistic Load Forecasting. *IEEE Trans. Smart Grid* **2018**, *9*, 1680–1687. [CrossRef]
4. Hong, T.; Fan, S. Probabilistic electric load forecasting: A tutorial review. *Int. J. Forecast.* **2016**, *32*, 914–938. [CrossRef]
5. Fan, S.; Hyndman, R.J. Short-Term Load Forecasting Based on a Semi-Parametric Additive Model. *IEEE Trans. Power Syst.* **2012**, *27*, 134–141. [CrossRef]
6. Nowotarski, J.; Liu, B.; Weron, R.; Hong, T. Improving short term load forecast accuracy via combining sister forecasts. *Energy* **2016**, *98*, 40–49. [CrossRef]
7. Hyndman, R.J.; Fan, S. Density Forecasting for Long-Term Peak Electricity Demand. *IEEE Trans. Power Syst.* **2010**, *25*, 1142–1153. [CrossRef]
8. Wang, Y.; Chen, Q.; Hong, T.; Kang, C. Review of Smart Meter Data Analytics: Applications, Methodologies, and Challenges. *IEEE Trans. Smart Grid* **2018**, in press. [CrossRef]
9. Quilumba, F.L.; Lee, W.-J.; Huang, H.; Wang, D.Y.; Szabados, R.L. Using Smart Meter Data to Improve the Accuracy of Intraday Load Forecasting Considering Customer Behavior Similarities. *IEEE Trans. Smart Grid* **2015**, *6*, 911–918. [CrossRef]

10. Xie, J.; Chen, Y.; Hong, T.; Laing, T.D. Relative Humidity for Load Forecasting Models. *IEEE Trans. Smart Grid* **2018**, *9*, 191–198. [CrossRef]
11. Xie, J.; Hong, T.; Xie, J.; Hong, T. Wind Speed for Load Forecasting Models. *Sustainability* **2017**, *9*, 795. [CrossRef]
12. Hong, T.; Wang, P.; Pahwa, A.; Gui, M.; Hsiang, S.M. Cost of temperature history data uncertainties in short term electric load forecasting. In Proceedings of the 2010 IEEE 11th International Conference on Probabilistic Methods Applied to Power Systems, Singapore, 14–17 June 2010; pp. 212–217.
13. Nedellec, R.; Cugliari, J.; Goude, Y. GEFCom2012: Electric load forecasting and backcasting with semi-parametric models. *Int. J. Forecast.* **2014**, *30*, 375–381. [CrossRef]
14. Hong, T.; Wang, P.; White, L. Weather station selection for electric load forecasting. *Int. J. Forecast.* **2015**, *31*, 286–295. [CrossRef]
15. Dordonnat, V.; Pichavant, A.; Pierrot, A. GEFCom2014 probabilistic electric load forecasting using time series and semi-parametric regression models. *Int. J. Forecast.* **2016**, *32*, 1005–1011. [CrossRef]
16. Charlton, N.; Singleton, C. A refined parametric model for short term load forecasting. *Int. J. Forecast.* **2014**, *30*, 364–368. [CrossRef]
17. Dordonnat, V.; Koopman, S.J.; Ooms, M.; Dessertaine, A.; Collet, J. An hourly periodic state space model for modelling French national electricity load. *Int. J. Forecast.* **2008**, *24*, 566–587. [CrossRef]
18. Lai, S.-H.; Hong, T. When One Size No Longer Fits All-Electric Load Forecasting with a Geographic Hierarchy. *SAS White Pap.* **2013**, 1–14. Available online: http://assets.fiercemarkets.net/public/sites/energy/reports/electricloadforecasting.pdf (accessed on 21 April 2019).
19. Hong, T.; Pinson, P.; Fan, S. Global Energy Forecasting Competition 2012. *Int. J. Forecast.* **2014**, *30*, 357–363. [CrossRef]
20. Wang, P.; Liu, B.; Hong, T. Electric load forecasting with recency effect: A big data approach. *Int. J. Forecast.* **2016**, *32*, 585–597. [CrossRef]
21. Fortin, F.-A.; Marc-André Gardner, U.; Parizeau, M.; Gagné, C. DEAP: Evolutionary Algorithms Made Easy François-Michel De Rainville. *J. Mach. Learn. Res.* **2012**, *13*, 2171–2175.

© 2019 by the authors. Licensee MDPI, Basel, Switzerland. This article is an open access article distributed under the terms and conditions of the Creative Commons Attribution (CC BY) license (http://creativecommons.org/licenses/by/4.0/).

Article

Averaging Predictive Distributions Across Calibration Windows for Day-Ahead Electricity Price Forecasting

Tomasz Serafin [1,2], Bartosz Uniejewski [1,2] and Rafał Weron [1,*]

[1] Department of Operations Research, Faculty of Computer Science and Management, Wrocław University of Science and Technology, 50-370 Wrocław, Poland
[2] Faculty of Pure and Applied Mathematics, Wrocław University of Science and Technology, 50-370 Wrocław, Poland
* Correspondence: rafal.weron@pwr.edu.pl; Tel.: +48-71-320-4525

Received: 19 May 2019; Accepted: 1 July 2019; Published: 3 July 2019

Abstract: The recent developments in combining point forecasts of day-ahead electricity prices across calibration windows have provided an extremely simple, yet a very efficient tool for improving predictive accuracy. Here, we consider two novel extensions of this concept to probabilistic forecasting: one based on Quantile Regression Averaging (QRA) applied to a set of point forecasts obtained for different calibration windows, the other on a technique dubbed Quantile Regression Machine (QRM), which first averages these point predictions, then applies quantile regression to the combined forecast. Once computed, we combine the probabilistic forecasts across calibration windows by averaging probabilities of the corresponding predictive distributions. Our results show that QRM is not only computationally more efficient, but also yields significantly more accurate distributional predictions, as measured by the aggregate pinball score and the test of conditional predictive ability. Moreover, combining probabilistic forecasts brings further significant accuracy gains.

Keywords: electricity price forecasting; predictive distribution; combining forecasts; average probability forecast; calibration window; autoregression; pinball score; conditional predictive ability

1. Introduction

After 25 years of intensive research, the *electricity price forecasting* (EPF) literature includes hundreds of publications, focused both on point [1,2] and probabilistic [3,4] predictions. However, very few studies try to find the optimal length of the calibration window or even consider calibration windows of different lengths. Instead, the typical approach has been to select ad-hoc a 'long enough' window, ranging from as few as 10 days to as much as five years. Only recently has this issue been tackled more systematically, initially by Hubicka et al. [5] and then in a follow up article by Marcjasz et al. [6]; note that the latter paper eventually appeared in print earlier than the original study.

Hubicka et al. [5] proposed a novel concept in energy forecasting that combined day-ahead predictions across different calibration windows ranging from 28 to 728 days. Using data from the Global Energy Forecasting Competition 2014, they showed that such *averaging across calibration windows* yielded better results than selecting ex-ante only one 'optimal' window length. They concluded that a mix of a few short- and a few long-term windows led to the best predictions. Marcjasz et al. [6] extended their analysis to other datasets and larger models. More importantly, they introduced a well-performing weighting scheme for averaging forecasts. Overall, their results confirmed earlier findings, but they advised to use slightly longer windows at the shorter end, especially when considering models with more explanatory variables (inputs). On the other hand, they concluded that including 3- instead of 2-year windows did not bring significant benefits. Marcjasz et al. recommended the **WAW**(56:28:112, 714:7:728) averaging scheme, i.e., past performance weighted combination of forecasts from six windows: 56-, 84-, 112-, 714-, 721- and 728-day long; we use Matlab notation to describe the sets of windows, e.g., (7:364) refers to all windows from 7 to 364 days, (14:7:105) to 14 window lengths: 14, 21, ..., 105 days, while (7, 364) to 7- and 364-day windows. In their empirical study, this averaging scheme performed very well and in most cases was not significantly outperformed by any other forecast.

Despite the innovative content, the above mentioned papers are limited to point predictions. To address this gap, here we consider two novel extensions of the *averaging across calibration windows* concept to probabilistic forecasting: one based on Quantile Regression Averaging (**QRA**) [7] and one using the Quantile Regression Machine (**QRM**) [8]. As the underlying statistical technique both use *quantile regression* [9], which has recently become the workhorse of probabilistic energy forecasting [10–14]. Moreover, both apply it to a pool of point forecasts obtained for calibration windows of different lengths and yield predictions for the 99 percentiles of the next day's price distribution for each hour. The difference between them lies in the choice of the regressors—**QRA** uses the point forecasts themselves, while **QRM** first averages them, then applies quantile regression to the combined forecast. Once computed, we combine the probabilistic forecasts across calibration windows by averaging probabilities of the corresponding predictive distributions, as in [15]. Although the latter works well in our study, the literature on combining predictive distributions offers alternatives [15–18] which could be considered as well.

Furthermore, since we want to focus on predictive distributions, we do not propose our own approach to point forecasting. Instead we select a well performing model and take its point forecasts as inputs to the **QRA** and **QRM** procedures. Our starting point is the study of Marcjasz et al. [6] and the autoregressive expert model they call **ARX2**, fitted to **asinh**-transformed day-ahead prices from two major power markets (Nord Pool and the PJM Interconnection) using one of the six suggested calibration window lengths for point forecasting (i.e., $T_0 = 56, 84, 112, 714, 721$ or 728 days). Next, we apply either **QRA** or **QRM** to these six series of point forecasts in a calibration window for probabilistic predictions of length $T = 14, 15, ..., 363$ or 364 days. It is important to emphasize that the calibration windows for point (T_0) and probabilistic (T) forecasts are two different objects—they may be of different length, are non-overlapping (the 'point' window directly precedes the 'probabilistic' one) and only the latter is evaluated in our study.

The remainder of the paper is structured as follows. In Section 2 we briefly describe the datasets. In Section 3 we first discuss the forecasting scheme, then recall the point forecasting setup of [6], in particular the **asinh** transformation and the **ARX2** model, and finally introduce our methodology for computing probabilistic predictions. In Section 4 we evaluate the obtained predictive distributions in terms of the Aggregate Pinball Score (APS) and test the conditional predictive ability using the approach of Giacomini and White [19]. Finally in Section 5 we wrap up the results and conclude.

2. Datasets

To evaluate our models we use datasets from two major power markets: the hydro-dominated and exhibiting strong seasonal variations Nord Pool (Northern Europe) and the world's largest competitive wholesale electricity market—the PJM Interconnection (Northeastern United States), with a balanced coal–gas–nuclear generation mix. Like in [6], the Nord Pool dataset comprises hourly system prices in EUR/MWh and day-ahead *consumption prognosis* for four Nordic countries (Denmark, Finland, Norway and Sweden), see Figure 1, while the PJM dataset—hourly prices and day-ahead load forecasts for the Commonwealth Edison (COMED) zone, see Figure 2. Note, however, that in our study both datasets start 364 days later, because—following the advice of Marcjasz et al. [6]—the longest calibration windows for point forecasts we consider are 728 days long. Consequently, the Nord Pool dataset spans 1674 days from 31 December 2013 to 31 July 2018 and the PJM dataset spans 1820 days from 9 April 2013 to 2 April 2018. Given that the longest calibration window for point predictions is $T_0 = 728$ days and for probabilistic forecasts $T = 364$ days, the out-of-sample test periods for evaluating probabilistic forecasts are: 27 December 2016 to 31 July 2018 (582 days) for Nord Pool and 5 April 2016 to 2 April 2018 (728 days) for PJM.

Energies **2019**, *12*, 2561

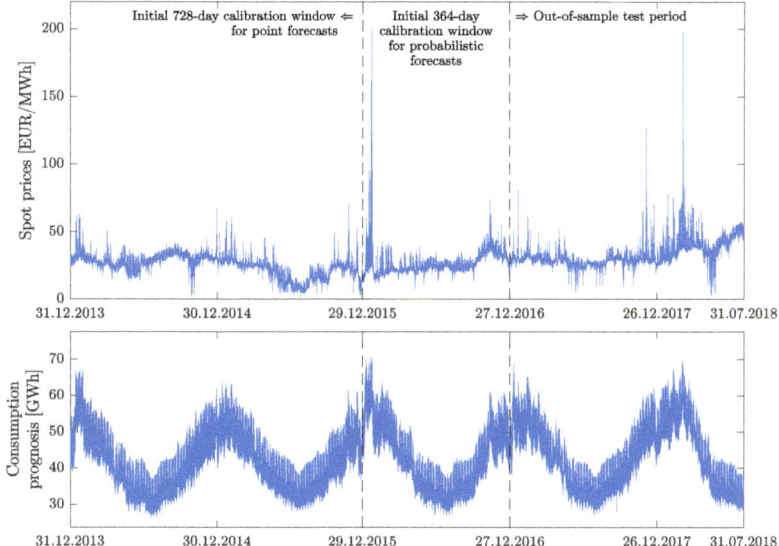

Figure 1. Nord Pool (NP) hourly system prices (*top*) and hourly consumption prognosis (*bottom*) from 31 December 2013 to 31 July 2018. The first dashed line marks the beginning (29 December 2015) of the initial 364-day calibration window for probabilistic forecasts, the second—the beginning (27 December 2016) of the 582-day long out-of-sample test period.

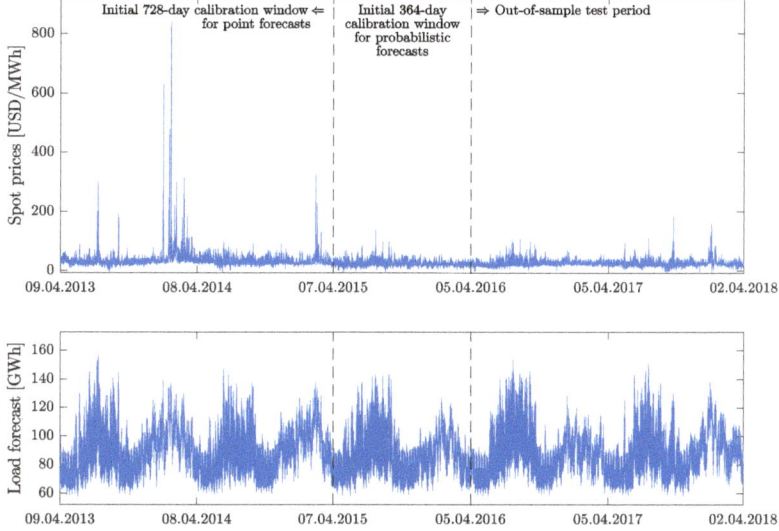

Figure 2. PJM hourly system prices (**top**) and hourly load forecasts (**bottom**) in the Commonwealth Edison (COMED) zone from 9 April 2013 to 2 April 2018. The first dashed line marks the beginning (7 April 2015) of the initial 364-day calibration window for probabilistic forecasts, the second—the beginning (5 April 2016) of the 728-day long out-of-sample test period.

3. Methodology

3.1. The Forecasting Scheme

In our empirical study, we use a rolling window scheme. Every day we compute 24 predictive distributions for each of the 24 hours of the next day, then move the calibration windows forward by one day and repeat the exercise. The calibration windows for probabilistic predictions range between $T = 14$ and 364 days, i.e., overall we consider 351 different window lengths. On the other hand, to obtain each predictive distribution we always use exactly six calibration windows for point forecasts: $T_0 = 56, 84, 112, 714, 721$ and 728 days.

Let us now illustrate this procedure using the Nord Pool dataset. To obtain predictive distributions for 27 December 2016 (denoted by an asterisk '*' in Figure 3) using the longest calibration window for probabilistic forecasts ($T = 364$ days; the light red shaded bar in the top part of Figure 3) we apply the **QRA** and **QRM** approaches to point forecasts obtained for the period from 29 December 2015 to 26 December 2016, i.e., the initial 364-day calibration window for probabilistic forecasts, see Figure 1. Of course, before we can do this, we have to compute the point forecasts for 29 December 2015 to 26 December 2016 by fitting the **ARX2** model to data in one of the six point forecasting calibration windows (blue shaded bars in the top part of Figure 3) directly preceding 29 December 2015, 30 December 2015, ..., 26 December 2016. For instance, for $T_0 = 728$ we use data in the initial 728-day window (i.e., 31 December 2013 to 28 December 2015; light blue bar), while for $T_0 = 56$ in a 56-day window from 11 November 2015 to 28 December 2015 (dark blue bar).

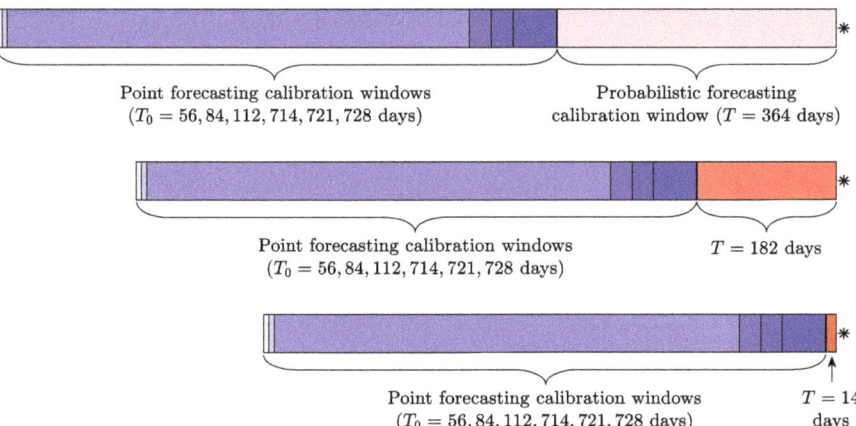

Figure 3. Illustration of our forecasting scheme. The target day for which the predictive distributions are computed is denoted by an asterisk '*'. The calibration windows for probabilistic forecasts (red shaded bars; the darker the shade the shorter the window) end on the previous day. Here three windows are plotted: the longest considered ($T = 364$ days; **top**), an intermediate ($T = 182$ days; **middle**) and the shortest considered ($T = 14$ days; **bottom**). They are directly preceded by six calibration windows for point forecasts (blue shaded bars; the darker the shade the shorter the window) of $T_0 = 56, 84, 112, 714, 721$ and 728 days.

Similarly, to obtain predictive distributions for 27 December 2016 using the shortest calibration window for probabilistic forecasts ($T = 14$ days; the dark red shaded bar in the bottom part of Figure 3) we apply the **QRA** and **QRM** approaches to point forecasts obtained for the 14-day period from 13 to 26 December 2016. Again, before we can do this, we have to compute the point forecasts for 13 December 2016 to 26 December 2016 by fitting the **ARX2** model to data in one of the six point forecasting calibration windows (blue shaded bars in the bottom part of Figure 3) directly preceding 13 December 2016, 14 December 2016, ..., 26 December 2016.

3.2. Computing Point Forecasts

Our point forecasting setup directly mimics that of Marcjasz et al. [6]. In particular, our modeling is conducted within a 'multivariate' framework, where we explicitly use the 'day × hour' matrix-like structure with $P_{d,h}$ representing the electricity price for day d and hour h. We calibrate our model to transformed data, i.e., $X_{d,h} = f(P_{d,h})$, where $f(\cdot)$ is the so-called *variance stabilizing transformation* (VST) [20]. In our case $f(\cdot)$ is the *area hyperbolic sine*:

$$X_{d,h} = \mathbf{asinh}(p_{d,h}) \equiv \log\left(p_{d,h} + \sqrt{p_{d,h}^2 + 1}\right), \tag{1}$$

where $p_{d,h} = \frac{1}{b}(P_{d,h} - a)$ are 'normalized' prices, a is the median of $P_{d,h}$ in the (point forecasting) calibration window and b is the sample *median absolute deviation* (MAD) around the sample median adjusted by a factor for asymptotically normal consistency to the standard deviation. This factor is $\frac{1}{z_{0.75}} \approx 1.4826$ where $z_{0.75}$ is the 75% quantile of the normal distribution. After computing the forecasts, we apply the inverse transformation, the *hyperbolic sine*, i.e., $p_{d,h} = \sinh(X_{d,h})$, in order to obtain the price predictions:

$$\widehat{P}_{d,h} = b \sinh(\widehat{X}_{d,h}) + a. \tag{2}$$

For computing point forecasts we use the well-performing $expert_{DoW,nl}$ model of Ziel and Weron [21], only expanded to include one exogenous variable (consumption or load forecast; see the bottom panels in Figures 1 and 2). Within this model, as in [6] denoted here by **ARX2**, the VST-transformed price on day d and hour h is given by:

$$X_{d,h} = \underbrace{\beta_{h,1} X_{d-1,h} + \beta_{h,2} X_{d-2,h} + \beta_{h,3} X_{d-7,h}}_{\text{autoregressive effects}} + \underbrace{\beta_{h,4} X_{d-1,min} + \beta_{h,5} X_{d-1,max}}_{\text{non-linear effects}}$$

$$+ \underbrace{\beta_{h,6} X_{d-1,24}}_{\text{midnight price}} + \underbrace{\beta_{h,7} C_{d,h}}_{\text{load forecast}} + \underbrace{\sum_{i=1}^{7} \beta_{h,7+i} D_i}_{\text{weekday dummies}} + \varepsilon_{d,h}. \tag{3}$$

where $X_{d-1,min}$ and $X_{d-1,max}$ are the minimum and the maximum of the previous day's 24 hourly prices, $X_{d-1,24}$ is the previous day's price at midnight (included in the model to take advantage of the influence it has on the prices for early morning hours [11,22]), $C_{d,h}$ is the known on day $d-1$ and VST-transformed consumption or load forecast for day d and hour h, and finally $D_1, ..., D_7$ are weekday dummies which capture the short-term seasonality. As in [6], the model parameters, i.e., $\beta_{h,1}, ..., \beta_{h,14}$, are estimated using Ordinary Least Squares (OLS), independently for each hour $h = 1, ..., 24$.

3.3. Computing Probabilistic Forecasts

In this paper, we introduce two extensions of the *averaging across calibration windows* concept to probabilistic forecasting. The first one is based on Quantile Regression Averaging (**QRA**) of Nowotarski and Weron [7], which has been found to perform very well in several test cases, including electricity price [3,11,23,24], load [25–27] and wind power forecasting [28]. Recall, that **QRA** involves applying quantile regression [9] to a pool of point forecasts. More precisely, the q-th quantile of the predicted variable (here: the electricity price $P_{d,h}$) is represented by a linear combination of predictor variables:

$$Q_q(P_{d,h}) = \mathbf{Y}_{d,h} \mathbf{w}_q, \tag{4}$$

where $\mathbf{w}_q$ is a vector of weights for quantile q, estimated by minimizing the so-called *pinball score* for each quantile, see Section 4. In our case, the predictor (or explanatory) variables are the point forecasts obtained for the six considered calibration windows:

$$\mathbf{Y}_{d,h} = \begin{bmatrix} 1 & \widehat{P}_{d,h}(56,T) & \widehat{P}_{d,h}(84,T) & \widehat{P}_{d,h}(112,T) & \widehat{P}_{d,h}(714,T) & \widehat{P}_{d,h}(721,T) & \widehat{P}_{d,h}(728,T) \end{bmatrix}, \tag{5}$$

where **1** denotes a $T \times 1$ vector of ones (i.e., the intercept) and $\widehat{\mathbf{P}}_{d,h}(\cdot, T)$ a $T \times 1$ vector of point forecasts, obtained for the **ARX2** model using one of the six calibration windows (i.e., $T_0 = 56, 84, 112, 714, 721$ or 728 days).

The second approach, after Marcjasz et al. [8] dubbed Quantile Regression Machine (**QRM**), first averages point predictions across the six calibration windows to yield $\overline{\mathbf{P}}_{d,h}(T)$, then applies quantile regression (4) to the combined forecast:

$$\mathbf{Y}_{d,h} = \begin{bmatrix} 1 & \overline{\mathbf{P}}_{d,h}(T) \end{bmatrix}. \tag{6}$$

Note, that $\overline{\mathbf{P}}_{d,h}(T)$ is of the same length as the six individual forecasts $\widehat{\mathbf{P}}_{d,h}(\cdot, T)$, hence the 'argument' T. The term Quantile Regression Machine is a compilation of 'quantile regression' and 'committee machine', since in [8] the authors considered committees of neural networks. Here, we use only regression models, but their weighted average across calibration windows reminds of the output of a committee machine. To reduce the complexity of our study we limit ourselves to the averaging scheme for point forecasts recommended by Marcjasz et al. [6], i.e., **WAW**(56:28:112, 714:7:728), which assigns weights based on yesterday's performance of the **ARX2** model in each calibration window, see Equation (5) in [6]. Regarding notation, recall from Section 3.1 that the calibration windows for probabilistic predictions range between $T = 14$ and 364 days. We use **QRA**(T) to denote a **QRA**-type and **QRM**(T) to denote a **QRM**-type probabilistic forecast obtained for a window of length T.

Finally, note that both for **QRA** and **QRM**, for each day d and hour h we forecast 99 percentiles, i.e., $q = 0.01, 0.02, ..., 0.99$, which approximate the entire predictive distribution relatively well. However, due to numerical inefficiency of quantile regression (4) the neighboring percentiles may be overlapping, leading to a phenomenon known as *quantile crossing* [29]. Hence, following Maciejowska and Nowotarski [11], the 99 quantiles are sorted to obtain monotonic quantile curves, independently for each day and hour.

3.4. Averaging Probabilistic Forecasts Across Calibration Windows

Given a set of probabilistic forecasts for calibration windows of $T = 14, 15, ..., 364$ days, we can combine all or some of them in one of two commonly used ways—by averaging probabilities or quantiles. The average quantile forecast, i.e., a horizontal average of the corresponding predictive distributions, is always more concentrated than the average probability forecast, i.e., a vertical average. While this feature is an advantage in many forecasting applications [15], it is not so in EPF [8,30]. Hence, in this study we only consider the *average probability forecast* defined as:

$$\frac{1}{n} \sum_{i=1}^{n} \hat{F}_i(x), \tag{7}$$

where $\hat{F}_i(x)$ is the i-th distributional forecast, actually a set of 99 percentiles obtained by setting $q = 0.01, 0.02, ..., 0.99$ in Equation (4), and n is the number of combined predictive distributions. We use this averaging scheme for both the **QRA**- and **QRM**-type probabilistic forecasts and respectively denote the combined forecasts by **QRA**($\mathcal{T}$) and **QRM**($\mathcal{T}$), where $\mathcal{T}$ is the vector of calibration windows used.

4. Results

We evaluate the probabilistic forecasts in terms of the Pinball Score, a so-called *proper* scoring rule and a special case of an asymmetric piecewise linear loss function [3,31]:

$$\mathbf{PS}\left(\widehat{Q}_q(P_{d,h}), P_{d,h}, q\right) = \begin{cases} (1-q)\left(\widehat{Q}_q(P_{d,h}) - P_{d,h}\right) & \text{for } P_{d,h} < \widehat{Q}_q(P_{d,h}), \\ q\left(P_{d,h} - \widehat{Q}_q(P_{d,h})\right) & \text{for } P_{d,h} \geq \widehat{Q}_q(P_{d,h}), \end{cases} \tag{8}$$

where $\widehat{Q}_q(P_{d,h})$ is the forecast of the q-th quantile of $P_{d,h}$ obtained using Equation (4), $P_{d,h}$ is the observed price and q is the quantile. The lower the score is, the more accurate are the probabilistic forecasts, i.e., the more concentrated are the predictive distributions. Note, that Equation (8) measures the predictive accuracy for only one particular quantile. However, it can be averaged across all percentiles (i.e., $q = 0.01, 0.02, ..., 0.99$) and all hours in the whole out-of-sample test period to yield the Aggregate Pinball Score (**APS**). Note, that computing the **APS** is equivalent to computing the quantile representation of the *Continuous Ranked Probability Score* (CRPS) [32,33], i.e., it is a discretization of the CRPS, which replaces an integral over all quantiles $q \in [0, 1]$ by a simpler to compute sum over 99 percentiles [3].

4.1. QRA vs. QRM

Let us first compare the two approaches to computing probabilistic forecasts described in Section 3.3. In Figure 4 we plot the **APS** across all 99 percentiles and all hours in the 582-day (for Nord Pool) and 728-day (for PJM) out-of-sample test periods, see Figures 1 and 2. Clearly, for each dataset and each T, **QRM**(T) yields more accurate probabilistic forecasts than **QRA**(T), with the results converging towards each other for larger windows. The latter may be due to too few observations for **QRA** to perform well for the shorter calibration windows, as it has 3.5 times more parameters to estimate than **QRM**. Given that computing predictive distributions via **QRM** is nearly three times faster than via **QRA**, we recommend the former approach. Note also, that initially we have considered calibration windows as short as 7 days. However, probabilistic forecasts for windows of less than 14 days perform poorly. For this reason they are not discussed in this paper.

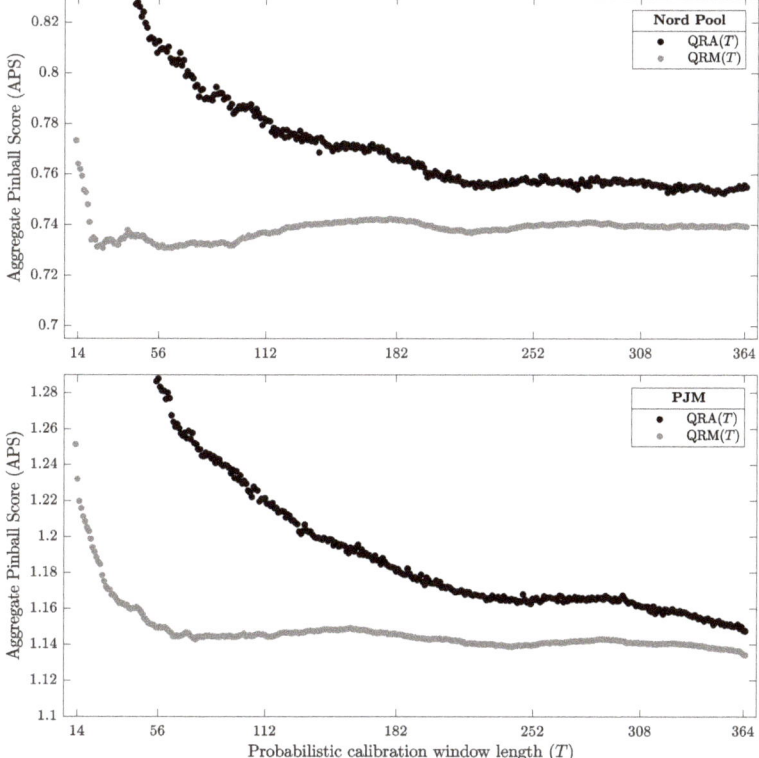

Figure 4. Aggregate Pinball Scores (APS) for the Nord Pool (**top**) and PJM (**bottom**) datasets as a function of $T = 14, 15, \ldots, 364$ days.

4.2. Averaging Probabilistic Forecasts across Calibration Windows

Now, let us turn to the core part of this empirical study. Analogously to [5,6], we examine several different combinations of calibration windows for probabilistic EPF. The results are illustrated in Figure 5 (for Nord Pool) and Figure 6 (for PJM). We can observe that for both datasets **QRA**(14:7:364), i.e., the combination across all window lengths with a step of 7 days depicted in both Figures by green triangles, is a top performer among **QRA**-type predictions. **QRM**(14:7:364) also yields good forecasts, that are more accurate than **QRM**(T) for all T, but is in turn outperformed by more sparse sets of windows. In particular, **QRA**(14:7:28,308:28:364) denoted by black stars and **QRA**(14:28:70,308:28:364) denoted by dark red squares outperform all other combinations. In addition, they are computationally efficient—they require computing probabilistic forecasts for only six calibration windows. This reminds of the results for point forecasts [5,6], where also combinations of three short and three long windows were top performers.

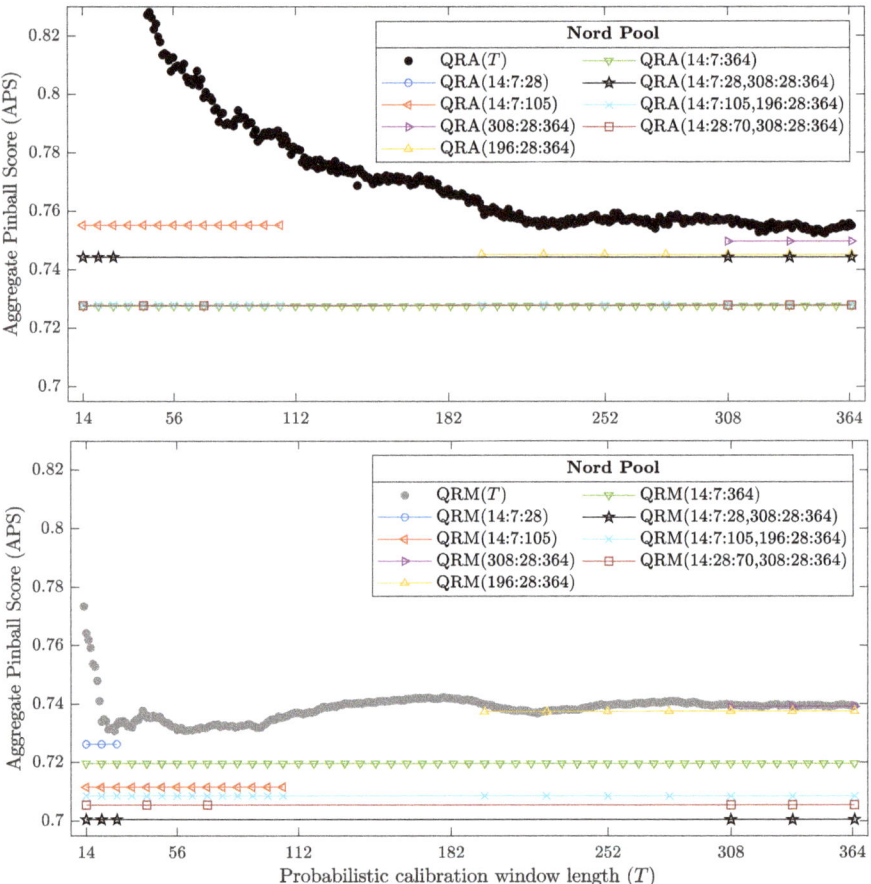

Figure 5. Aggregate pinball scores (APS) for probabilistic QRA (**top**) and QRM (**bottom**) forecasts for the Nord Pool dataset. Filled circles refer to individual probabilistic calibration window lengths and lines with symbols indicate window lengths selected for averaging forecasts.

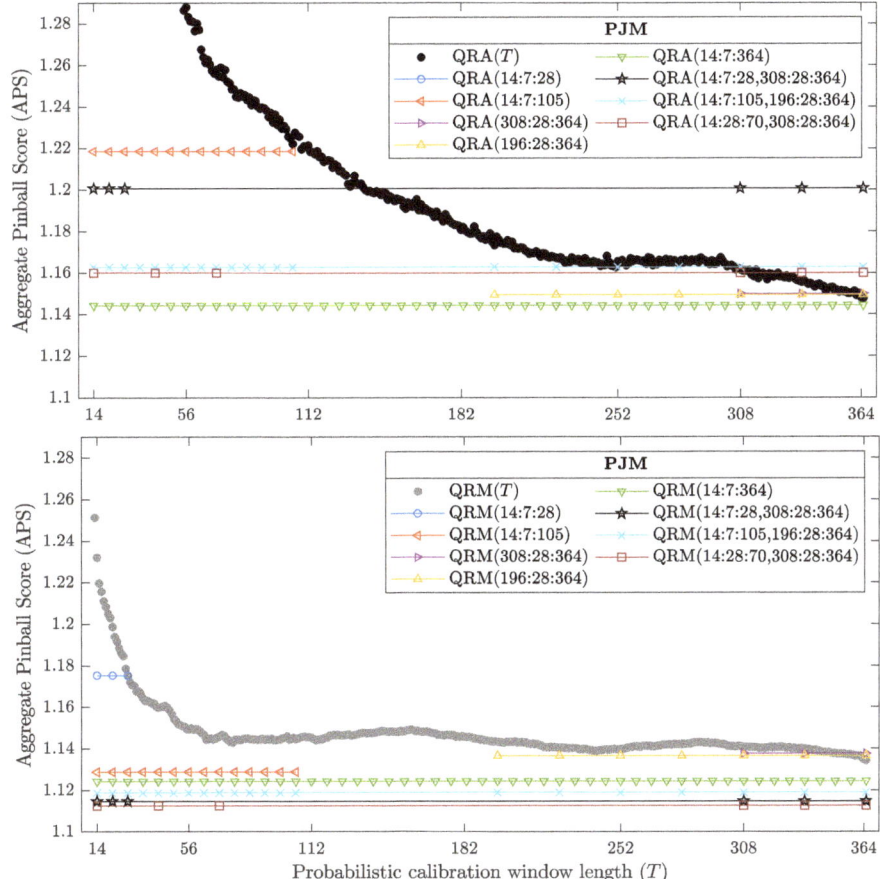

Figure 6. Aggregate pinball scores (APS) for probabilistic QRA (**top**) and QRM (**bottom**) forecasts for the PJM dataset. Filled circles refer to individual probabilistic calibration window lengths and lines with symbols indicate window lengths selected for averaging forecasts.

4.3. The CPA Test and Statistical Significance

The analyzed so far Aggregate Pinball Scores (APS) can be used to provide a ranking, but do not allow drawing statistically significant conclusions on the outperformance of the forecasts of one window set by those of another. Therefore, we use the Giacomini and White [19] test for *conditional predictive ability* (CPA), which can be regarded as a generalization of the commonly used Diebold-Mariano test for *unconditional* predictive ability. Since only the CPA test accounts for parameter estimation uncertainty, it is the preferred option. Here, one statistic for each pair of window sets is computed based on the 24-dimensional vector of Pinball Scores for each day:

$$\Delta_{X,Y,d} = \|\mathbf{PS}_{X,d}\| - \|\mathbf{PS}_{Y,d}\|, \tag{9}$$

where $\|\mathbf{PS}_{\mathcal{T},d}\| = \sum_{h=1}^{24} |\mathbf{PS}_{d,h}|$ for window set $\mathcal{T}$. For each pair of window sets and each dataset we compute the *p*-value of the CPA test with null $H_0 : \boldsymbol{\phi} = 0$ in the regression [19]:

$$\Delta_{X,Y,d} = \boldsymbol{\phi}' \mathbb{X}_{d-1} + \varepsilon_d, \tag{10}$$

where $\mathbb{X}_{d-1}$ contains elements from the information set on day $d-1$, i.e., a constant and lags of $\Delta_{X,Y,d}$.

In Figure 7 we illustrate the obtained p-values using 'chessboards', analogously as in [20,21,30,34] for the Diebold-Mariano test, i.e., we use a heat map to indicate the range of the p-values—the closer they are to zero ($\to$ dark green) the more significant is the difference between the forecasts of a window set on the X-axis (better) and the forecasts of a window set on the Y-axis (worse). Evidently, the CPA test results confirm and emphasize the observations made in Section 4.2. In particular, **QRM**(14:7:28,308:28:364) and **QRM**(14:28:70,308:28:364) significantly outperform all other window sets; additionally for Nord Pool data **QRM**(14:7:28,308:28:364) is significantly more accurate than **QRM**(14:28:70,308:28:364). Comparing the averaging schemes, it is worth noting that in no case does **QRA** significantly outperform the corresponding **QRM** scheme. On the other hand, the opposite can be observed for several cases for both datasets. This result reinforces our recommendation of using the **QRM** scheme for probabilistic EPF, regardless of the number of predictions combined.

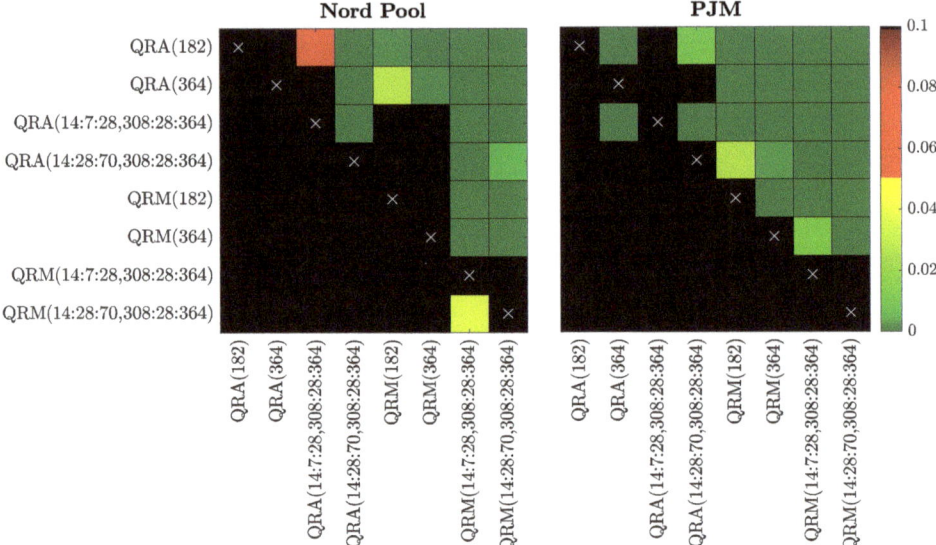

Figure 7. Results of the conditional predictive ability (CPA) test [19] for forecasts of selected models for the Nord Pool (*left*) and PJM (*right*) data. We use a heat map to indicate the range of the p-values—the closer they are to zero ($\to$ dark green) the more significant is the difference between the forecasts of a model on the X-axis (better) and the forecasts of a model on the Y-axis (worse).

5. Conclusions

In this paper, we take the *averaging across calibration windows* concept to a new level. Motivated by the results of Hubicka et al. [5] and Marcjasz et al. [6] for point forecasts, we consider two extensions of this approach to probabilistic forecasting: one based on Quantile Regression Averaging (**QRA**) [7], the other on the Quantile Regression Machine (**QRM**) [8]. Both methods apply quantile regression to a pool of point forecasts in order to obtain predictions for the 99 percentiles of the next day's price distribution. The difference between them lies in the choice of the regressors—**QRA** uses the point forecasts themselves, while **QRM** the combined point forecast.

Somewhat surprisingly, it turns out that it is not only more computationally efficient, but also better in terms of the Pinball Score to first average point predictions and then apply quantile regression to the combined forecast, than to use quantile regression directly on the individual point forecasts. In other words, a more general approach (**QRA**) is outperformed by a two-step technique (**QRM**). We believe that this outcome is due to two factors: the simpler model structure with fewer parameters to estimate and the more accurate point forecasts used as inputs. As Uniejewski et al. [30] have recently shown, more accurate point forecasts directly translate into better probabilistic forecasts computed via quantile regression. In addition, averaging point forecasts for a few adequately chosen calibration window lengths leads to decreasing the MAE by about 5% compared to the best point forecast obtained for a single window, even selected *ex-post* [5,6].

Regarding the selection of calibration windows for combining forecasts, similarly to the results for point predictions, the best performing combinations are those averaging a small number of short- and long-term windows. In particular, **QRM**(14:7:28,308:28:364) significantly outperforms all other considered window sets and is recommended for averaging probabilistic forecasts. We should emphasize that in this study we are using only one, relatively simple way of combining probabilistic forecasts, i.e., the *average probability forecast*, but the literature on combining predictive distributions offers interesting alternatives [15–18]. They could be tested in this context as well. Moreover, the proposed methodology can be easily extended to other areas of energy forecasting (e.g., load, solar, wind) or other types of markets (e.g., intraday). Finally, the economic benefits from using more accurate probabilistic electricity price predictions could be evaluated, as is becoming more common in the point forecasting literature [35,36].

Author Contributions: Conceptualization, R.W.; Investigation, T.S. and B.U.; Software, T.S. and B.U.; Validation, R.W.; Writing—original draft, T.S. and B.U.; Writing—review & editing, R.W.

Funding: This work was partially supported by the National Science Center (NCN, Poland) through grant No. 2015/17/B/HS4/00334 (to T.S. and R.W.) and No. 2016/23/G/HS4/01005 (to B.U.).

Conflicts of Interest: The authors declare no conflict of interest.

References

1. Weron, R. Electricity price forecasting: A review of the state-of-the-art with a look into the future. *Int. J. Forecast.* **2014**, *30*, 1030–1081. [CrossRef]
2. Lago, J.; De Ridder, F.; De Schutter, B. Forecasting spot electricity prices: Deep learning approaches and empirical comparison of traditional algorithms. *Appl. Energy* **2018**, *221*, 386–405. [CrossRef]
3. Nowotarski, J.; Weron, R. Recent advances in electricity price forecasting: A review of probabilistic forecasting. *Renew. Sustain. Energy Rev.* **2018**, *81*, 1548–1568. [CrossRef]
4. Ziel, F.; Steinert, R. Probabilistic mid- and long-term electricity price forecasting. *Renew. Sustain. Energy Rev.* **2018**, *94*, 251–266. [CrossRef]
5. Hubicka, K.; Marcjasz, G.; Weron, R. A note on averaging day-ahead electricity price forecasts across calibration windows. *IEEE Trans. Sustain. Energy* **2019**, *10*, 321–323. [CrossRef]
6. Marcjasz, G.; Serafin, T.; Weron, R. Selection of calibration windows for day-ahead electricity price forecasting. *Energies* **2018**, *11*, 2364. [CrossRef]
7. Nowotarski, J.; Weron, R. Computing electricity spot price prediction intervals using quantile regression and forecast averaging. *Comput. Stat.* **2015**, *30*, 791–803. [CrossRef]
8. Marcjasz, G.; Uniejewski, B.; Weron, R. Probabilistic electricity price forecasting with NARX networks: Combine point or probabilistic forecasts? *Int. J. Forecast.* **2019**, forthcoming.
9. Koenker, R.W. *Quantile Regression*; Cambridge University Press: Cambridge, UK, 2005.
10. Juban, R.; Ohlsson, H.; Maasoumy, M.; Poirier, L.; Kolter, J. A multiple quantile regression approach to the wind, solar, and price tracks of GEFCom2014. *Int. J. Forecast.* **2016**, *32*, 1094–1102. [CrossRef]
11. Maciejowska, K.; Nowotarski, J. A hybrid model for GEFCom2014 probabilistic electricity price forecasting. *Int. J. Forecast.* **2016**, *32*, 1051–1056. [CrossRef]
12. Andrade, J.; Filipe, J.; Reis, M.; Bessa, R. Probabilistic price forecasting for day-ahead and intraday markets: Beyond the statistical model. *Sustainability* **2017**, *9*, 1990. [CrossRef]
13. Bracale, A.; Carpinelli, G.; De Falco, P. Developing and comparing different strategies for combining probabilistic photovoltaic power forecasts in an ensemble method. *Energies* **2019**, *12*, 1011. [CrossRef]
14. Ziel, F. Quantile regression for the qualifying match of GEFCom2017 probabilistic load forecasting. *Int. J. Forecast.* **2019**. [CrossRef]
15. Lichtendahl, K.C.; Grushka-Cockayne, Y.; Winkler, R.L. Is it better to average probabilities or quantiles? *Manag. Sci.* **2013**, *59*, 1594–1611. [CrossRef]

16. Gneiting, T.; Ranjan, R. Combining predictive distributions. *Electron. J. Stat.* **2013**, *7*, 1747–1782. [CrossRef]
17. Bassetti, F.; Casarin, R.; Ravazzolo, F. Bayesian nonparametric calibration and combination of predictive distributions. *J. Am. Stat. Assoc.* **2018**, *113*, 675–685. [CrossRef]
18. Baran, S.; Lerch, S. Combining predictive distributions for the statistical post-processing of ensemble forecasts. *Int. J. Forecast.* **2018**, *34*, 477–496. [CrossRef]
19. Giacomini, R.; White, H. Tests of conditional predictive ability. *Econometrica* **2006**, *74*, 1545–1578. [CrossRef]
20. Uniejewski, B.; Weron, R.; Ziel, F. Variance stabilizing transformations for electricity spot price forecasting. *IEEE Trans. Power Syst.* **2018**, *33*, 2219–2229. [CrossRef]
21. Ziel, F.; Weron, R. Day-ahead electricity price forecasting with high-dimensional structures: Univariate vs. multivariate modeling frameworks. *Energy Econ.* **2018**, *70*, 396–420. [CrossRef]
22. Ziel, F. Forecasting electricity spot prices using LASSO: On capturing the autoregressive intraday structure. *IEEE Trans. Power Syst.* **2016**, *31*, 4977–4987. [CrossRef]
23. Gaillard, P.; Goude, Y.; Nedellec, R. Additive models and robust aggregation for GEFCom2014 probabilistic electric load and electricity price forecasting. *Int. J. Forecast.* **2016**, *32*, 1038–1050. [CrossRef]
24. Kostrzewski, M.; Kostrzewska, J. Probabilistic electricity price forecasting with Bayesian stochastic volatility models. *Energy Econ.* **2019**, *80*, 610–620. [CrossRef]
25. Liu, B.; Nowotarski, J.; Hong, T.; Weron, R. Probabilistic load forecasting via Quantile Regression Averaging on sister forecasts. *IEEE Trans. Smart Grid* **2017**, *8*, 730–737. [CrossRef]
26. Sigauke, C.; Nemukula, M.; Maposa, D. Probabilistic hourly load forecasting using additive quantile regression models. *Energies* **2018**, *11*, 2208. [CrossRef]
27. Wang, Y.; Zhang, N.; Tan, Y.; Hong, T.; Kirschen, D.; Kang, C. Combining probabilistic load forecasts. *IEEE Trans. Smart Grid* **2019**. [CrossRef]
28. Zhang, Y.; Liu, K.; Qin, L.; An, X. Deterministic and probabilistic interval prediction for short-term wind power generation based on variational mode decomposition and machine learning methods. *Energy Convers. Manag.* **2016**, *112*, 208–219. [CrossRef]
29. Chernozhukov, V.; Fernandez-Val, I.; Galichon, A. Quantile and probability curves without crossing. *Econometrica* **2010**, *73*, 1093–1125.
30. Uniejewski, B.; Marcjasz, G.; Weron, R. On the importance of the long-term seasonal component in day-ahead electricity price forecasting. Part II—Probabilistic forecasting. *Energy Econ.* **2019**, *79*, 171–182. [CrossRef]
31. Gneiting, T. Quantiles as optimal point forecasts. *Int. J. Forecast.* **2011**, *27*, 197–207. [CrossRef]
32. Gneiting, T.; Ranjan, R. Comparing density forecasts using thresholdand quantile-weighted scoring rules. *J. Bus. Econ. Stat.* **2011**, *29*, 411–422. [CrossRef]
33. Laio, F.; Tamea, S. Verification tools for probabilistic forecasts of continuous hydrological variables. *Hydrol. Earth Syst. Sci.* **2007**, *11*, 1267–1277. [CrossRef]
34. Uniejewski, B.; Weron, R. Efficient forecasting of electricity spot prices with expert and LASSO models. *Energies* **2018**, *11*, 2039. [CrossRef]
35. Kath, C.; Ziel, F. The value of forecasts: Quantifying the economic gains of accurate quarter-hourly electricity price forecasts. *Energy Econ.* **2018**, *76*, 411–423. [CrossRef]
36. Maciejowska, K.; Nitka, W.; Weron, T. Day-ahead vs. Intraday—Forecasting the price spread to maximize economic benefits. *Energies* **2019**, *12*, 631. [CrossRef]

© 2019 by the authors. Licensee MDPI, Basel, Switzerland. This article is an open access article distributed under the terms and conditions of the Creative Commons Attribution (CC BY) license (http://creativecommons.org/licenses/by/4.0/).

Article

Solar Power Interval Prediction via Lower and Upper Bound Estimation with a New Model Initialization Approach

**Peng Li [1], Chen Zhang [2] and Huan Long [2],*

[1] State Grid Zhejiang Electrical Power Research Institute, Hangzhou 310014, China; lipeng_hz@163.com
[2] School of Electrical Engineering, Southeast University, Nanjing 211189, China; czhangmail@163.com
* Correspondence: hlong@seu.edu.cn

Received: 29 September 2019; Accepted: 28 October 2019; Published: 30 October 2019

Abstract: This paper proposes a new model initialization approach for solar power prediction interval based on the lower and upper bound estimation (LUBE) structure. The linear regression interval estimation (LRIE) was first used to initialize the prediction interval and the extreme learning machine auto encoder (ELM-AE) is then employed to initialize the input weight matrix of the LUBE. Based on the initialized prediction interval and input weight matrix, the output weight matrix of the LUBE could be obtained, which was close to optimal values. The heuristic algorithm was employed to train the LUBE prediction model due to the invalidation of the traditional training approach. The proposed model initialization approach was compared with the point prediction initialization and random initialization approaches. To validate its performance, four heuristic algorithms, including particle swarm optimization (PSO), simulated annealing (SA), harmony search (HS), and differential evolution (DE), were introduced. Based on the experiment results, the proposed model initialization approach with different heuristic algorithms was better than the point prediction initialization and random initialization approaches. The PSO can obtain the best efficiency and effectiveness of the optimal solution searching in four heuristic algorithms. Besides, the ELM-AE can weaken the over-fitting phenomenon of the training model, which is brought in by the heuristic algorithm, and guarantee the model stable output.

Keywords: solar power prediction; interval prediction; lower and upper bound estimation; extreme learning machine; heuristic algorithm

1. Introduction

With the increasing global energy consumption, renewable energy and its application technologies have received extensive attention and are being studied enthusiastically. The intermittent nature and volatility of renewable energy, as significant factors, restrict its exploitation and penetration. An accurate forecast is required to guarantee the stability and economy of power systems. However, the randomness and indeterminacy of natural resources bring great difficulties for solar power predictions.

Traditional solar power point prediction provides limited forecast information, which causes risk [1]. Solar power interval prediction offering interval information under a certain confidence level breaks a new pathway to handle forecasting uncertainty. The interval prediction technology aims at predicting a narrow interval, encompassing as many predicted points as possible. The high-quality prediction intervals are of benefit to static safety analysis and risk evaluation in power systems. However, solar power interval prediction attracts less attention compared to point prediction. The existing prominent interval prediction methods include the statistical method and data-driven method.

The statistical methods are first employed to construct the prediction interval. Statistical methods usually require prior knowledge or distribution assumption of forecasting errors [2–5]. They often assume that the forecast errors follow a normal distribution with a zero mean or t-student distribution [6]. The bootstrap [7], Bayesian [8], mean-variance estimation [5], and delta methods [9] are the four prominent and traditional methods. These four methods were analyzed from calculations, interval precision, and interval width, which revealed that each method had its shortcomings [10]. The prediction errors display different characters and differ in various application fields. Thus, it is important to make the appropriate distribution assumption, which might result in poor forecasting performance. Li et al. acquired a precise distribution characteristic based on the divided dataset by the envelope-based clustering algorithm. There are also several statistical methods without any prior assumption for probabilistic prediction, such as kernel density estimation [11], ensemble simulations [12], and quantile regression [3].

Data-driven methods are gradually introduced to avoid distribution assumptions. The lower and upper bound estimation (LUBE) structure for interval prediction was first developed by Khosravi et al. [13]. Two output units of the neural network (NN) model were employed to represent the upper and lower bounds of the predicted interval. Such nonparametric models are further widely utilized in many research works [14–16]. In the process of training the LUBE, two prominent evaluation metrics, coverage probability and interval width, are considered. Due to their contradictoriness, the LUBE training can be considered as a multi-objective or single-objective optimization model [17–19]. In [20], a new multi-objective optimization method using multi-objective swarm algorithm was proposed to adjust the machine learning model, which revealed superior forecasting performance to the single-objective one. In [21], the Pareto optimal solutions were used to construct a multi-objective framework and Pareto solutions obtained an ensemble of optimal solutions. Due to the discontinuous differentiability of the cost function, it is hard to train the NN through the traditional analytical algorithm. Heuristic algorithms such as particle swarm optimization (PSO) and simulated annealing (SA) are employed in this situation.

Most previous interval prediction methods based on LUBE models concentrate on the building of optimization objective and the selection of intelligent algorithms. The initialization method of the NN parameters is rarely studied. However, the initial solution of heuristic algorithms significantly affects their evolution process and performance.

ELM-AE employed in this paper aims at enhancing the generalization capability of the forecasting model. Besides this, current application objects of interval prediction mainly include wind speed, wind power, electricity load, and electricity price prediction. Solar energy, as a representative renewable resource, also deserves some discussion for interval prediction.

This paper proposes a new model initialization approach for the prediction interval based on the LUBE structure. The ELM-AE is first utilized to initialize the input weight matrix of the LUBE model and the linear regression interval estimation (LRIE) is then used to initialize the prediction interval. The initial prediction interval obtained by LRIE is then employed to update the initial parameters of the LUBE model. Numerous comparison experiments are conducted to validate the performance of the proposed model initialization approach.

Some experiments using the proposed initialization approach, traditional initialization approach, and random initialization approach are implemented with the same sample data. Different heuristic algorithms, including particle swarm optimization (PSO), simulated annealing (SA), harmony search (HS), and differential evolution (DE), are conducted to evaluate the impact of the initial solution on different heuristic algorithms.

The remainder of this paper is organized as follows. Section 2 introduces the LUBE method employing ELM and two primary evaluation indices of forecasting intervals. The proposed model initialization approach is described in Section 3. Experiments and results are reviewed in Section 4. Finally, Section 5 makes some conclusions of this work and discusses some guidelines for future work.

2. Lower and Upper Bound Estimation

The LUBE method utilizing the neural network structure has been widely used to estimate the prediction interval. The schematic diagram of the LUBE method is shown in Figure 1. The ELM with two output nodes is regarded as the prediction model of LUBE. The output of the two output nodes represents the predicted upper and lower bound. Because the actual predicted interval is unknown and uncertain, the traditional background propagation algorithm cannot be used to train the ELM. The training of the ELM is converted to a parameter optimization problem and the heuristic algorithm is utilized to obtain the optimal parameters of the LUBE.

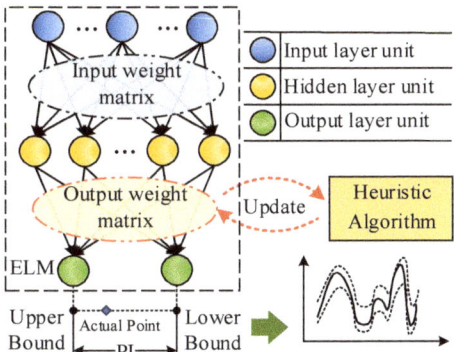

Figure 1. The schematic diagram of the lower and upper bound estimation (LUBE).

2.1. ELM

The ELM introduced by Huang, et al. [22] is a single-hidden layer feed forward neural network with excellent generalization performance and fast learning speed. Thus, the ELM is utilized as the prediction model in this work. In Figure 1, the ELM only has three layers, the input layer, hidden layer, and output layer. Two neuron units in the output layer separately represent the upper and lower bounds of the predicted interval.

In the normal ELM model, suppose that N samples $\{x_j, t_j\}_{j=1}^{N}$ are given, where $x_j \in R^n$, representing the input vector, $t_j \in R^m$, representing the target vector. The input data are transmitted to the L dimensional feature space constructed by the hidden layer and the output element of the network is obtained by Equation (1):

$$f_L(\mathbf{x}) = \sum_{i=1}^{L} \beta_i \mathbf{h}_i(\mathbf{x}) = \mathbf{h}(\mathbf{x})\boldsymbol{\beta} \quad (1)$$

where $\mathbf{h}(x)$ indicates the outputs of hidden neuron node and the L element corresponds to the outputs of L hidden nodes generated from activation function. Likewise, $\boldsymbol{\beta} = [\beta_1, \ldots, \beta_L]^T$ is the output weight matrix. The goal of the single hidden layer neural network is to minimize the error between the output value and the actual quantity. In matrix form, the target of the network is achieved by Equation (2):

$$\mathbf{H}\boldsymbol{\beta} = \mathbf{T} \quad (2)$$

where $\mathbf{H} = [\mathbf{h}^T(\mathbf{x}_1), \ldots, \mathbf{h}^T(\mathbf{x}_N)]^T$ and $\mathbf{T} = [t_1, \ldots, t_N]^T$. Thus, in ELM, the output weight β can be expressed as Equation (3):

$$\hat{\boldsymbol{\beta}} = \mathbf{H}^\dagger \mathbf{T} \quad (3)$$

where $\mathbf{H}^\dagger$ is the Moore–Penrose generalized inverse of matrix.

2.2. The Evaluation and Training of LUBE

To evaluate the prediction performance, the mean prediction interval width, PI_{mean}, and prediction interval coverage probability (PICP) in (4)–(5) are introduced. The PI_{mean} qualifies the width of prediction interval. The PICP indicates the percentage of the probability targets covered by the corresponding prediction intervals.

$$\text{PI}_{\text{mean}} = \frac{1}{N}\sum_{i=1}^{N}(\bar{t}_i - \underline{t}_i) \qquad (4)$$

$$\text{PICP} = \frac{1}{N}\sum_{i=1}^{N} 1_{[\underline{t}_i, \bar{t}_i]}(t_i) \qquad (5)$$

where $\bar{t}_i$ and $\underline{t}_i$ are the predicted upper and lower bounds of the dataset $\{(x_i, t_i), i = 1, \ldots, N\}$. Since the forecasting interval width is strongly associated with the range of the targets, normalized width evaluation index is more suitable for intuitional comparison. A new normalized index, called prediction interval normalized root-mean-square width (PINRW), is employed as in (6) [14]:

$$\text{PINRW} = \frac{1}{R}\sqrt{\frac{1}{N}\sum_{i=1}^{N}(\bar{t}_i - \underline{t}_i)^2} \qquad (6)$$

where R is the range of the forecasting targets. In general, R is equal to the difference between the maximum and minimum values of the training set.

The PI_{mean} and PICP (or PINRW) are the contradictory indexes. An ideal interval aims to maximize PICP and minimize PI_{mean} simultaneously. However, a balance and a compromise are required in practice. The cost function coverage width-based criterion (CWC) is introduced to evaluate the predicted interval. The flexible index combines prediction interval coverage percentage and width simultaneously, which could evaluate the overall performance of the prediction intervals and guide the generation of intervals:

$$\text{CWC} = \text{PINRW}(1 + 1_{[0,\delta)}(\text{PICP})e^{-\eta(\text{PICP}-\delta)}) \qquad (7)$$

where the hyper-parameter η magnifies the difference between PICP and δ, which should be a large value.

The training of LUBE can be regarded as an optimization problem. The minimization of CWC is the optimization objective and the output weight matrix of ELM is the independent variable. The heuristic algorithm is employed to obtain the optimal output weight matrix by minimizing CWC. The initialization of the output weight matrix can be generated randomly, called the random initialization (RI) approach. The output weight matrix obtained by the point prediction approach also can be utilized to initialize the output weight matrix of the LUBE, called the point initialization (PI) approach [13].

3. Proposed Model Initialization Approach

In the traditional LUBE interval prediction model, the random input weight matrix and search capacity of the heuristic algorithm significantly impact the final prediction performance. In this section, the proposed model initialization approach is introduced, including prediction interval initialization and input weight matrix initialization, shown in Figure 2. The initial prediction interval $\{\mathbf{T}^U, \mathbf{T}^L\}^0$ was first obtained by the prediction interval width initialization method. The input weight matrix $\boldsymbol{\beta}^T$ was then generated by the ELM-AE. The initial output weight matrix w_0 was finally gained by training the LUBE prediction model based on the initial prediction interval and input weight matrix.

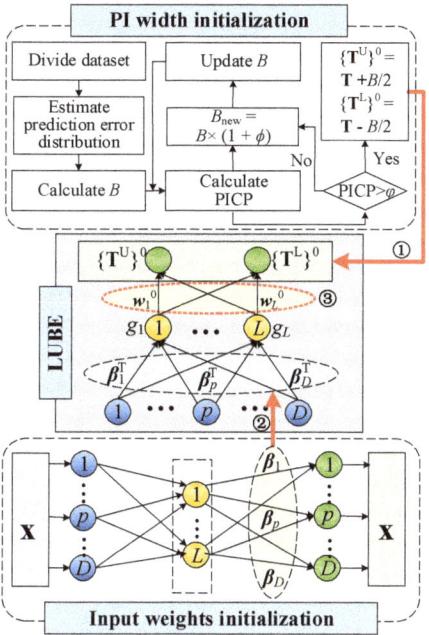

Figure 2. The model initialization approach.

3.1. Prediction Interval Initialization

In order to initialize the interval width and estimate initial prediction interval value of the whole training dataset $\{\mathbf{T}^U, \mathbf{T}^L\}^0$, cross-validation technology was utilized. In Figure 2, the training dataset $\{\mathbf{X}, \mathbf{T}\}$ is first divided for cross validation. In each part, suppose $\mathbf{T} = \mathbf{X}\boldsymbol{\Phi} + \boldsymbol{\mu}$, $E(\boldsymbol{\mu}) = 0$ and $Var(\boldsymbol{\mu}) = \sigma^2 \mathbf{I}$. Then, the prediction error e_0 on a single future observation $\{\mathbf{X}_0, \mathbf{T}_0\}$ follows the normal distribution $e_0 \sim N(0, \sigma^2(1+\mathbf{X}_0(\mathbf{X}^T\mathbf{X})^{-1}\mathbf{X}_0^T))$ shown in (8) and (9):

$$E(e_0) = E\left(\mu_0 - \mathbf{X}_0((\mathbf{X}^T\mathbf{X})^{-1}\mathbf{X}^T(\mathbf{X}\boldsymbol{\Phi}+\boldsymbol{\mu})-\boldsymbol{\Phi})\right) \\ = E(\mu_0 - \mathbf{X}_0(\mathbf{X}^T\mathbf{X})^{-1}\mathbf{X}^T\boldsymbol{\mu}) = 0 \tag{8}$$

$$Var(e_0) = Var\left(\mu_0 - \mathbf{X}_0((\mathbf{X}^T\mathbf{X})^{-1}\mathbf{X}^T\boldsymbol{\mu}\right) \\ = \sigma^2\left(1 - \mathbf{X}_0(\mathbf{X}^T\mathbf{X})^{-1}\mathbf{X}_0^T\right) \tag{9}$$

Therefore, the prediction interval is $T'_0 \pm t_{\alpha/2, n-m}\sigma'\sqrt{\left(1+\mathbf{X}_0(\mathbf{X}^T\mathbf{X})^{-1}\mathbf{X}_0^T\right)}$. According to (4), the initial interval width is B_0.

The PI_{mean} of the $\{\mathbf{X}, \mathbf{T}\}$ is then calculated as the initial interval width, denoted as B_0. To guarantee the expected prediction interval coverage probability φ is satisfied, the B_0 should be further adjusted through the binary search algorithm [18]. The actual value of the target $\mathbf{T}$ and the initial interval width B compose the initial prediction interval, $\{\mathbf{T}^U, \mathbf{T}^L\}^0$, shown as (10):

$$\{\mathbf{T}^U\}^0 = \mathbf{T}+B/2, \{\mathbf{T}^L\}^0 = \mathbf{T}-B/2 \tag{10}$$

The details of prediction interval width initialization are presented in the following steps (see Algorithm 1):

Algorithm 1 Prediction Interval Width Initialization

Input:
Training data $\{X,T\} = \{(x_i, t_i) | x_i \in \Re,\ t_i \in \Re,\ i = 1, 2, \ldots, n\}$;
Nominal confidence α;
Number of data subsets m;
Expected prediction interval coverage probability φ.
Output:
Initial Prediction Interval $\{T^U, T^L\}^0$.
(a) Calculate initial interval width B_0 of $\{X, T\}$.
(a-1) Divide the training dataset $\{X, T\}$ into m subsets;
(a-2) Sequentially select one subset as the testing data and other subsets are regarded as the training data;
(a-3) Separately estimate the prediction error distribution and prediction interval of each test data according to α based on the corresponding training data.
(a-4) Calculate PI_{mean} by (4), denoted as B_0.
(a-5) Calculate $\{T^U, T^L\}^0$ by (10), where $B = B_0$.
(b) Calculate the PICP of the ELM trained through $\{T^U, T^L\}^0$ for the training set. If PICP $< \varphi$, go to (c). If PICP $\geq \varphi$, output $\{T^U, T^L\}^0$.
(c) Update B by the binary search algorithm
(c-1) $B_{new} = B \times (1 + \phi)$;
(c-2) Update $\{T^U, T^L\}^0$ by (10) and go to (b), where $B = B_{new}$;

3.2. Input Weight Matrix Initialization

In conventional training of the ELM model, its input weights are randomly generated. However, the random input weights have influence on the output weights training, which further impact the prediction performance, especially model training through the heuristic algorithm.

The ELM-AE is capable of learning a useful feature representation [23], which could improve the generalization of the predicted model via projecting the input data into a different dimensional space [24]. ELM-AE has shown good capacity to learn a useful feature representation. The unique differentiation of the specific input data is reduced by the feature transformation. The generalization of the predicted model will be improved via projecting the input data into a different dimensional space.

In ELM-AE, the output data were the same as the input data shown Figure 3. The output weight β represents the information transformation from the feature space to input data. The steps of initializing input weights of ELM through ELM-AE are described in Algorithm 2.

Algorithm 2 Input Weight Initialization of LUBE

Input:
Training dataset $\{X\} = \{x_i | x_i \in \Re,\ i = 1, 2, \ldots, n\}$;
The number of hidden layer nodes of ELM-AE L.
Output:
Input weight matrix of LUBE
(a) Randomly generate the input weight matrix **a** and bias vector **b** of the ELM-AE hidden nodes.
(b) Orthogonalize **a** and **b**:
$a^T a = I,\ b^T b = 1$
(c) Calculate the output of ELM-AE hidden nodes **H**
$H = [g(a_l, b_l, x_i)]_{i=1,\ldots,n,\ l=1,\ldots,L}$
(d) Calculate output weight β of ELM-AE, and the input matrix of LUBE is β^T
$\beta = \left(\frac{I}{C} + H^T H\right)^{-1} H^T X$

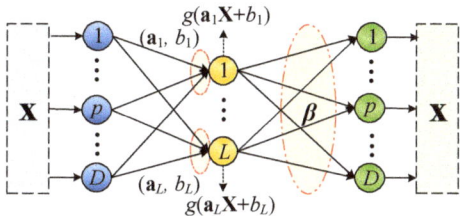

Figure 3. The structure of ELM-AE.

4. Experiment and Results

The bi-hourly solar power data utilized in this paper were collected from a grid-connected photovoltaic (PV) system over two years, from 1 July 2010 to 16 June 2012. The PV system was installed on the rooftop of an academic building located in the Coloane island of Macau. The related two-year data recorded by environmental detector and PV power monitoring in real-time were employed to validate the methods. The data included the date, time, solar radiation, temperature, wind speed, and solar power. In the interval prediction model, the historical time series data of the solar power, P_{t-2} and P_{t-1}, and weather data were generated as input variables to predict P_t. One-step-ahead prediction was carried out in this section. The majority of the data (70%) were regarded as the training dataset, while the rest were the test dataset.

4.1. Parameter Settings

To evaluate the proposed LUBE interval prediction model, several widely used heuristic algorithms, including PSO, DE, SA, and HS, were utilized. The PSO algorithm developed by Kennedy and Eberhart [25] was applied to various fields for its strong convergence performance. The DE algorithm combining the genetic algorithm evolution mechanism with the crossover and mutation operation evolves the population, and DE is suitable to handle non differentiable, such as discrete, problems [26]. SA can accept the worse solution to replace the current optimum by the probabilistic technique, which contributes to high search capacity in a large solution space [27]. HS is a simple meta-heuristic algorithm originated by the improvisation process of jazz musicians, which has been strongly criticized as a special case of the well-established evolution strategies algorithm [28].

The parameter settings of four heuristic algorithms are shown as Table 1. In PSO, the inertia weight linearly decreased from 0.7 to 0.1 in the iteration process. In DE, the crossover constant decreased linearly from 0.3 to 0.1 as the iteration increased. In HS, the pitch adjusting rate and bandwidth descended linearly within the range of (0.05, 1) and (1, 50). These algorithms with different characteristics would require different maximum iteration time for an anticipant result. The PSO, which is good at local optimum could converge within a fewer number of iterations. However, SA and HS, as global optimum algorithms, require more iterations to optimize intensively. Thus, the maximum iteration times of the PSO, SA, HS, and DE algorithms re set to 500, 10,000, 2500, and 500, respectively.

In ELM, the number of hidden layer neurons and the tradeoff parameter C were set to 188 and 512 through the point prediction and 5-fold cross-validation technique.

Considering the slight difference of the training and test data, the δ of CWC used equal to 93% in the training set and 90% in the test set, separately. The η was selected as 50 to greatly penalize prediction intervals with a coverage probability lower than δ. In order to leave a certain margin of optimization and avoid being trapped in local optimum, the expected PICP, φ, was set at 95%.

The experiments with different initialization approaches and heuristic algorithms were conducted. Each case was repeated five times to reduce the randomness influence of the dataset partitioning and heuristic algorithms. All experiments in this paper were implemented on a personal notebook computer with i5-4210U CPU and the 8 GB memory.

Table 1. Parameter settings of heuristic algorithms.

Algorithms	Parameter	Value
PSO	Particle size	100
	Inertia weight	(0.1, 0.7)
	Cognitive acceleration constant	1.5
	Social acceleration constant	2.5
DE	Population size	100
	Scaling factor F	0.005
	Crossover parameter CR	(0.1, 0.3)
SA	Initial temperature	5
	Re-annealing interval	50
	Cooling factor	0.9
HS	Harmony memory size	25
	Harmony memory considering rate	0.98
	Pitch adjusting rate	(0.05, 0.1)
	Bandwidth	(1, 50)

4.2. Computational Results

In the experiments, the width initialization, point initialization, and random initialization approaches were abbreviated as WI, PI and RI. The terms w/ ELM-AE and w/o ELM-AE mean the initialization approach with ELM-AE and without ELM-AE, respectively.

Tables 2–5 summarize the average and worst values of the different cases w/ and w/o ELM-AE. Due to the randomness character of the heuristic algorithm, it obtained different results for each optimization, so the result of the average and worst cases can have a comprehensive understanding of the performance and robustness of the algorithms. In Tables 2 and 3, the average case of HS for PI obtained a CWC of 49.36%, but the worst result was 66.4%. In Tables 4 and 5, the average case of SA for WI acquired a CWC of 67.86%, but the worst result was 145.29%, which was almost twice as much as the average case. The model combining PSO with WI w/ ELM-AE produced the best and the most stable prediction result among all the cases.

The training accuracy of WI and PI was similar in Table 2, but the WI behaved more stably than PI in the aspect of the test set. In general, The WI was superior to the PI and RI. The RI performed the worst in all the cases.

Comparing Table 2 with Table 3, the initialization approach with ELM-AE was better than the one without ELM-AE. The ELM-AE can significantly improve the prediction performance of RI in both of the training and test datasets. In PI and WI, the CWC of the training set with ELM-AE was higher than the one without ELM-AE. However, the performance of the test set was the reverse. It is implied that the ELM-AE can reduce the over-fitting phenomenon in the training process, and improve the stability of the test set by impairing the random impact of initial weight.

Table 2. Comparison of average results of different cases with ELM-AE.

AVERAGE		WI w/ELM-AE			PI w/ELM-AE			RI w/ELM-AE		
		CWC	PICP	PINRW	CWC	PICP	PINRW	CWC	PICP	PINRW
PSO	Training	26.64%	93.01%	26.64%	26.47%	93.01%	26.47%	117.57%	93.01%	117.57%
	Test	25.89%	91.30%	25.89%	35.70%	90.90%	25.95%	114.58%	94.14%	114.58%
SA	Training	27.91%	93.06%	27.91%	28.31%	93.07%	28.31%	269.35%	93.03%	269.35%
	Test	35.99%	91.15%	26.78%	28.68%	91.91%	28.68%	272.57%	93.86%	272.57%
HS	Training	32.75%	94.68%	32.75%	48.46%	95.64%	48.46%	476.17%	96.26%	476.17%
	Test	32.75%	93.47%	32.75%	49.36%	94.89%	49.36%	615.22%	95.36%	462.11%
DE	Training	32.75%	94.68%	32.75%	37.22%	94.68%	37.22%	350.1%	98.11%	350.10%
	Test	32.75%	93.47%	32.75%	37.47%	93.37%	37.47%	330.66%	98.03%	330.66%

Table 3. Comparison of the worst results of different cases with ELM-AE.

WORST		WI w/ ELM-AE			PI w/ ELM-AE			RI w/ ELM-AE		
		CWC	PICP	PINRW	CWC	PICP	PINRW	CWC	PICP	PINRW
PSO	Training	28.19%	93.01%	28.19%	25.94%	93.01%	25.94%	66.04%	91.08%	66.04%
	Test	28.28%	92.50%	28.28%	50.35%	89.97%	24.98%	206.21%	93.01%	206.21%
SA	Training	28.11%	93.03%	28.11%	28.88%	93.16%	28.88%	328.42%	93.03%	328.42%
	Test	27.54%	91.88%	27.54%	29.91%	92.32%	29.91%	337.88%	94.32%	337.88%
HS	Training	34.37%	95.04%	34.37%	66.69%	98.27%	66.69%	371.12%	93.41%	371.12%
	Test	34.37%	94.05%	34.37%	66.40%	98.76%	66.40%	908.87%	88.55%	296.42%
DE	Training	34.37%	95.04%	34.37%	37.22%	94.68%	37.22%	429.44%	99.28%	429.44%
	Test	34.37%	94.05%	34.37%	37.47%	93.37%	37.47%	417.74%	98.09%	417.74%

Table 4. Comparison of average results of different cases without ELM-AE.

AVERAGE		WI w/o ELM-AE			PI w/o ELM-AE			RI w/o ELM-AE		
		CWC	PICP	PINRW	CWC	PICP	PINRW	CWC	PICP	PINRW
PSO	Training	22.66%	93.01%	22.36%	24.55%	93.01%	24.55%	199.31%	93.01%	199.31%
	Test	50.70%	89.43%	22.56%	51.99%	89.64%	24.17%	347.39%	91.99%	187.52%
SA	Training	26.36%	93.07%	26.36%	27.07%	93.16%	27.07%	697.12%	93.04%	697.12%
	Test	67.86%	89.17%	25.34%	43.98%	90.50%	27.61%	669.95%	93.58%	669.95%
HS	Training	30.48%	94.65%	30.48%	102.92%	92.75%	33.21%	994.26%	96.43%	994.26%
	Test	30.56%	92.77%	30.56%	42.28%	91.89%	34.26%	1037.78%	97.38%	1037.78%
DE	Training	26.30%	93.30%	26.30%	24.86%	93.38%	24.86%	688.64%	96.36%	688.64%
	Test	30.57%	90.71%	25.77%	83.97%	89.22%	23.72%	641.13%	94.58%	641.13%

Table 5. Comparison of the worst results of different cases without ELM-AE.

WORST		WI w/o ELM-AE			PI w/o ELM-AE			RI w/o ELM-AE		
		CWC	PICP	PINRW	CWC	PICP	PINRW	CWC	PICP	PINRW
PSO	Training	21.79%	93.01%	21.79%	23.48%	93.01%	23.48%	195.77%	93.01%	195.77%
	Test	77.47%	88.06%	21.30%	82.94%	87.88%	21.36%	620.14%	87.88%	159.74%
SA	Training	25.25%	93.04%	25.25%	26.08%	93.06%	26.08%	845.49%	93.06%	845.49%
	Test	145.29%	86.73%	23.69%	71.09%	88.86%	25.67%	889.78%	93.12%	889.78%
HS	Training	31.25%	95.02%	31.25%	29.88%	93.27%	29.88%	1143.80%	97.97%	1143.80%
	Test	31.25%	93.16%	31.25%	69.00%	89.35%	28.92%	1273.26%	99.64%	1273.26%
DE	Training	22.45%	93.12%	22.45%	30.29%	93.04%	30.29%	825.84%	97.04%	825.84%
	Test	44.69%	89.70%	20.69%	228.93%	86.11%	28.61%	791.94%	96.14%	791.94%

4.2.1. Result Analysis 1—Initialization Approach

The prediction interval results employing different initialization approaches with ELM-AE and PSO are shown in Figures 4–6. It is clear that most actual power points can be covered in the interval due to the expected PICP equal to 0.93.

In the enlarged views of Figures 4 and 5, both of the predicted boundaries of WI and PI can accurately trace the fluctuation of the power curve and preform similarly.

However, in the turning points, such as the 8th and 20th points in the left view and the 6th and 18th points in the right view, the predicted interval of WI was narrower than PI. Thus, the whole predicted interval of WI was more uniform than PI and its predicted result was better, which is in accordance with Table 2.

In Tables 2–5, for the average test result, the best PINRW for RI was 114.58%, while the worst result for PI and WI was 49.36%. The PINRW of RI was much larger than WI and PI. Thus, the predicted interval of RI intends to employ a universal upper and lower limit to cover as many points as possible, as shown in Figure 6, which has no guidance function.

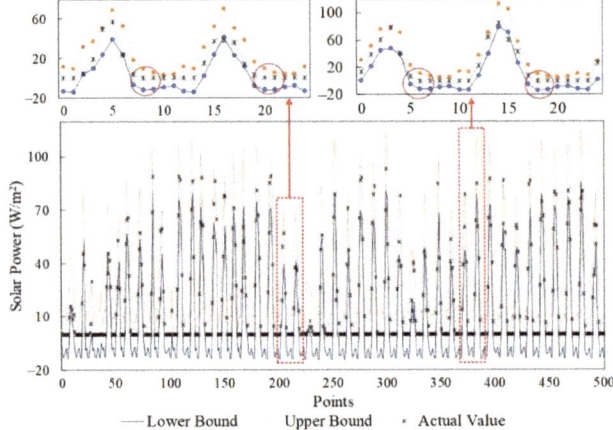

Figure 4. The prediction results obtained by PSO and WI with ELM-AE.

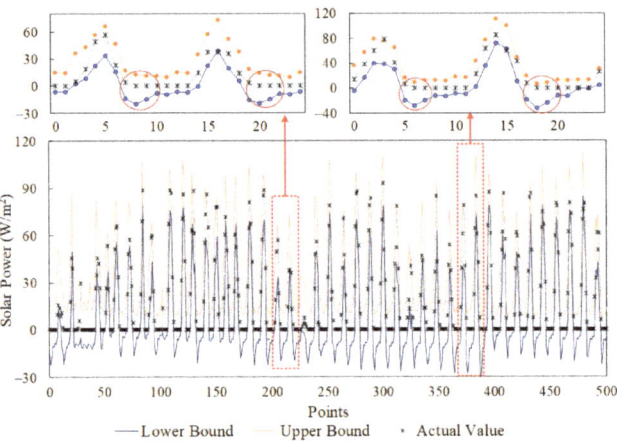

Figure 5. The prediction results obtained by PSO and PI with ELM-AE.

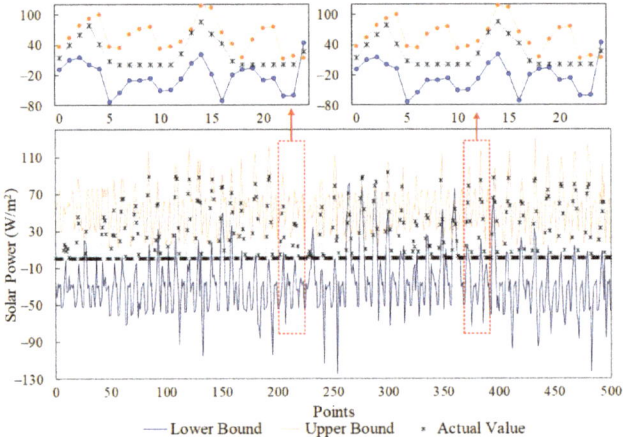

Figure 6. The prediction result obtained by PSO and RI with ELM-AE.

The CWC convergence curves of PSO for different cases are shown as representative in Figure 7. It is apparent that the CWC initial values in RI were significantly larger than other non-random initialization ways. The curves of RI almost converged around 250 iterations. The WI and PI could achieve stable values around 100 iterations. Besides, the converged value of RI was much larger than the WI and PI. Thus, it is concluded that the RI is not a good choice of LUBE initialization.

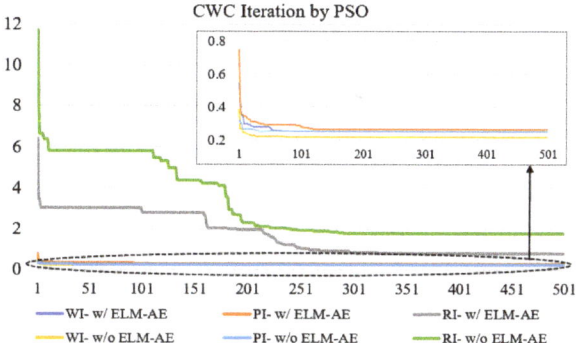

Figure 7. CWC of the best solution in training process for PSO.

4.2.2. Result Analysis 2—ELM-AE

Figures 8–10 display the predicted intervals by WI, PI, and RI w/o ELM-AE. Comparing Figures 4 and 5 with Figures 8 and 9, it is apparent that no matter whether or not the WI and PI utilized ELM-AE, their performances were generally close. In the enlarged views, the predicted interval of WI and PI w/o ELM-AE was narrower than the one with ELM-AE, especially points in the night. This is because the ELM-AE impaired the randomness impact of LUBE, which also reduced the diversity of the solutions and further impacted the optimal solution evolution. Thus, the initialization approach w/o ELM-AE had a higher chance of obtaining the global optimal solution than the one w/ ELM-AE, but it also caused unstable performances due to the over-fitting phenomenon.

When ELM-AE was not utilized in RI in Figure 10, the performance dropped drastically, resulting in the fluctuation range of interval reaching ±200. Thus, the employment of ELM-AE can facilitate RI by reducing the divergence of the model.

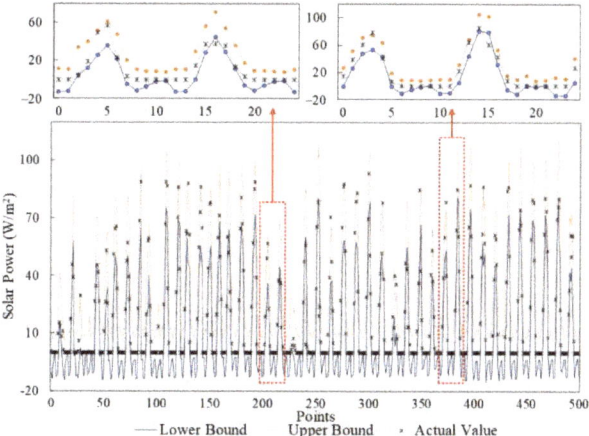

Figure 8. The prediction results obtained by PSO and WI w/o ELM-AE.

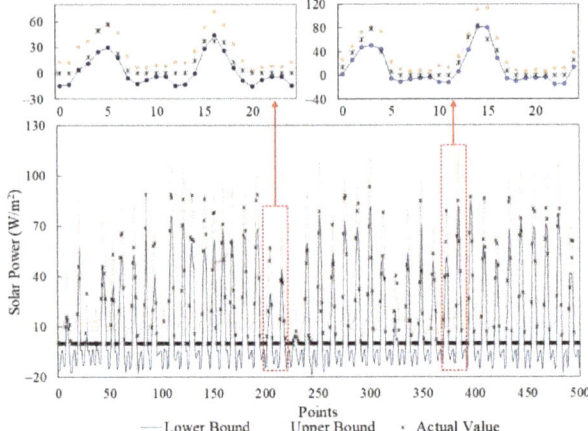

Figure 9. The prediction result obtained by PSO and PI w/o ELM-AE.

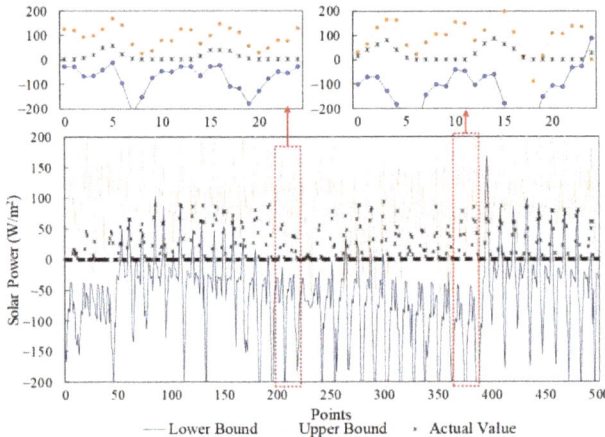

Figure 10. The prediction result obtained by PSO and RI w/o ELM-AE.

To clearly explain the role of ELM-AE, the characteristics of the input weight matrix of ELM in LUBE was analyzed in detail. The rank of the input weight matrix was not influenced. The mean absolute value of the input weight matrix grew down from 0.5014 to 0.1146 and the matrix sparsity dropped from 0.2451 to 0.1368 after adding ELM-AE. Thus, the ELM-AE displays the role of the feature extraction and weakens the overfitting of the trained model.

4.2.3. Result Analysis 3—Heuristic Algorithm

To display performances of different heuristic algorithms, the prediction intervals through the WI w/ ELM-AE model optimized by SA and HS are shown in Figures 11 and 12. It is obvious that the lower bounds in Figures 11 and 12 are lower than the one in Figure 4, resulting in the wider prediction interval. The PSO preformed the best among all the heuristic algorithms. In theory, the SA, HS, and DE algorithms have better global optimum search capacity than PSO. However, in the case of LUBE prediction interval, their evolutionary efficiencies were too low and could not obtain a good result in the limited computational time. In Tables 2 and 4, the prediction results of the HS and DE are the same for WI. This is because their optimal solutions, obtained in the initialization, stayed the same in the whole progress due to their low evolutionary efficiency.

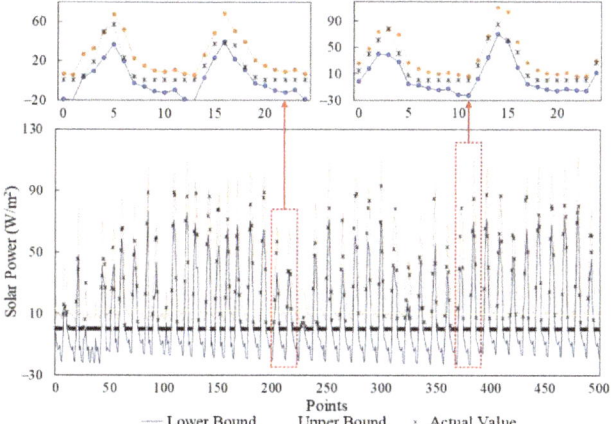

Figure 11. The prediction results obtained by SA and WI w/ ELM-AE.

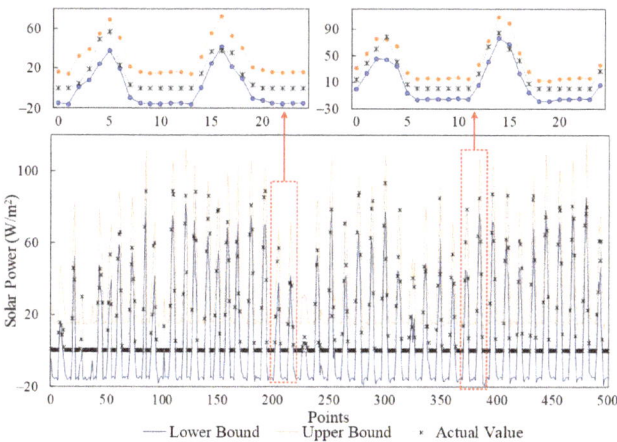

Figure 12. The prediction result obtained by HS with WI w/ ELM-AE.

Figures 13 and 14 display the predicted interval optimized by SA and HS based on WI w/o ELM-AE. Combined with Tables 3 and 5, the trained LUBE prediction model displayed obvious over-fitting phenomenon. The PSO ha the most serious over-fitting phenomenon among all the heuristic algorithms due to its good capacity of solving optimization problems.

The iterative time for various heuristic algorithms is another factor affecting the model performance, especially for online prediction. Average computational times for different heuristic algorithms and initialization approaches are shown in Tables 6 and 7. The training time is directly impacted by the evaluation times of the cost function. The evaluation times of PSO, SA, HS, and DE were 50,000, 50,000, 62,500, and 50,000, respectively. The running time of PSO and DE was close and the SA cost the most computational time. Comparing Table 6 with Table 7, it is obvious that the experiments without ELM-AE ran longer than the ones with ELM-AE in all cases. This is because the ELM-AE makes the input weight matrix of the LUBE sparse, which reduces the computational load and cuts down the time.

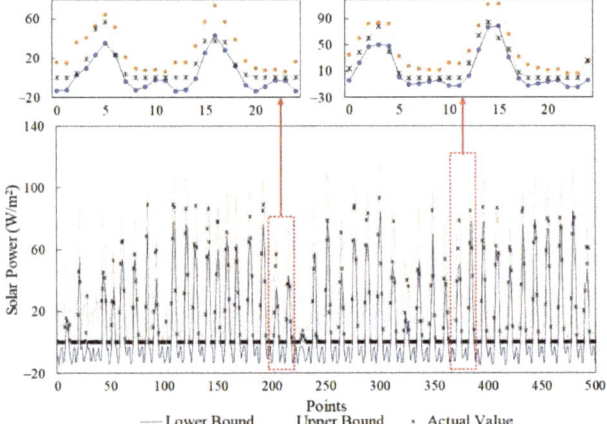

Figure 13. The prediction result obtained by SA and WI w/o ELM-AE.

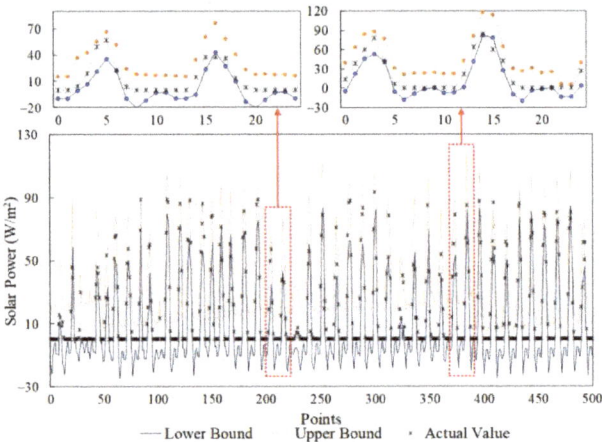

Figure 14. The prediction result obtained by HS and WI w/o ELM-AE.

Table 6. Average computational time (s) for different initialization approaches with ELM-AE.

Algorithms	WI	PI	RI
PSO	1374.22	1369.52	1360.97
SA	2235.66	2224.44	2223.50
HS	1670.62	1647.88	1647.88
DE	1398.42	1367.05	1366.08

Table 7. Average computational time (s) for different initialization approaches without ELM-AE.

Algorithms	WI	PI	RI
PSO	1533.88	1673.48	1484.03
SA	2327.96	2322.45	2330.64
HS	1647.88	1795.76	1788.86
DE	1474.17	1469.56	1465.95

5. Conclusions

Renewable energy generation forecasting technology contributes to decreasing the uncertainty and randomness of renewable resources and can provide essential reference information for the scheduling and operation of the power system. Interval prediction with a statistical confidence level is good at quantifying the uncertainties of the forecasting power. This paper proposed a new LUBE interval prediction framework based on the point prediction technology of ELM. The ELM-AE was employed to generate input weight matrix β^T; then PI width initialization way acquired the initial output weight matrix w_0, satisfying the presupposed PICP. Finally, the output weights of ELM were further optimized through a heuristic algorithm. Four algorithms, PSO, DE, SA, and HS, were implemented to verify the performance of the proposed mechanism. Different experimental settings were combined into different contrast experiments to validate and analyze the impacts of different settings on the model performance.

The prediction performance of WI was slightly superior to the property of PI generally. At some power curve turning points, WI could more reasonably constrain the prediction interval and avoid a large prediction margin. The simulation experiments revealed that ELM-AE could significantly decrease the matrix sparsity and the mean absolute value of the input weight matrix, which are statistically equal to 0.5 when the matrix is randomly generated from a uniform random distribution between (−1, 1). The over-fitting of the learned model was weakened and the generalization ability of the model improved when using ELM-AE. The PSO algorithm achieved the best prediction performance among the four algorithms under various situations. The SA, HS, and DE algorithms performed poorly in the limited computational time, and the HS and DE algorithms could hardly further optimize the output weight matrix. The performance of the model was also constrained by the limitations of the heuristic algorithms and was related to the algorithm parameters. However, the PSO resulted in the most severe over-fitting phenomenon for a sharp prediction interval. In general, the proposed LUBE model with a new model initialization approach would acquire a faithful prediction interval with more detailed optimization and stable generalization performance.

Although the LUBE approach can forecast the interval covering the solar power accurately, the width of prediction intervals at different times of day was consistent. However, it is apparent that the power value is zero in the night and that the nighttime interval can be narrower. The mechanism of LUBE makes the width of interval in different periods consistent, which deserves improvement in further research. Some normal optimization technique for neural networks also can be added to the prediction model framework to improve the learning performance, such as the ensemble learning of multiple neural networks. The evaluation fitness function transforms the original multi-objective problem into a single-objective problem for simplification. CWC could effectively guarantee the PICP of prediction intervals, but the penalty term also restricts and intervenes in the search for an optimal solution, which results in some feasible solutions being unavailable. In the future, it is expected to explore a new evaluation mechanism that could systematically balance the coverage probability and the width of the prediction interval.

Acronyms

CWC	Coverage width-based criterion
DE	Differential evolution
ELM-AE	ELM auto encoder
HS	Harmony search
LRIE	linear regression interval estimation
LUBE	Lower and upper bound estimation
NN	Neural network
PI	Point initialization approach
PV	Photovoltaic
PICP	Prediction interval coverage probability

PINRW   PI normalized root-mean-square width
PSO   Particle swarm optimization
RI   Random initialization approach
SA   Simulated annealing
WI   Width initialization approach
w/ ELM-AE   Initialization approach with ELM-AE
w/o ELM-AE   Initialization approach without ELM-AE
ELM   Extreme learning Machine

Author Contributions: H.L. supervised the project and designed the experiment. Resources and data curation, P.L.; writing—original draft preparation, C.Z.; writing—review and editing, H.L.

Funding: This research was funded by the Science and Technology Program of State Grid Corporation of Zhejiang Province under Grand 5211DS17001Z, the National Natural Science Foundation of China under Grant 51807023, Natural Science Foundation of Jiangsu Province under Grant BK20180382.

Conflicts of Interest: The authors declare no conflict of interest.

References

1. Khosravi, A.; Mazloumi, E.; Nahavandi, S.; Creighton, D.; Lint, J.W.C.V. Prediction intervals to account for uncertainties in travel time prediction. *IEEE Trans. Intell. Transp. Syst.* **2011**, *12*, 537–547. [CrossRef]
2. Saez, D.; Avila, F.; Olivares, D.; Canizares, C.; Marin, L. Fuzzy prediction interval models for forecasting renewable resources and loads in microgrids. *IEEE Trans. Smart Grid* **2015**, *6*, 548–556. [CrossRef]
3. He, Y.; Liu, R.; Li, H.; Wang, S.; Lu, X. Short-term power load probability density forecasting method using kernel-based support vector quantile regression and copula theory. *Appl. Energy* **2017**, *185 Pt 1*, 254–266. [CrossRef]
4. Tahmasebifar, R.; Sheikh-El-Eslami, M.K.; Kheirollahi, R. Point and interval forecasting of real-time and day-ahead electricity prices by a novel hybrid approach. *IET Gener. Transm. Distrib.* **2017**, *11*, 2173–2183. [CrossRef]
5. Yang, X.; Ma, X.; Kang, N.; Maihemuti, M. Probability interval prediction of wind power based on kde method with rough sets and weighted markov chain. *IEEE Access* **2018**, *6*, 51556–51565. [CrossRef]
6. Yun, S.L.; Scholtes, S. Empirical prediction intervals revisited. *Int. J. Forecast.* **2014**, *30*, 217–234.
7. Sheng, C.; Zhao, J.; Wang, W.; Leung, H. Prediction intervals for a noisy nonlinear time series based on a bootstrapping reservoir computing network ensemble. *IEEE Trans. Neural Netw. Learn. Syst.* **2013**, *24*, 1036–1048. [CrossRef]
8. MacKay, D.J.C. The evidence framework applied to classification networks. *Neural Comput.* **1992**, *4*, 720–736. [CrossRef]
9. Veaux, R.D.D.; Schumi, J.; Ungar, S.L.H. Prediction intervals for neural networks via nonlinear regression. *Technometrics* **1998**, *40*, 273–282. [CrossRef]
10. Kothari, S.C.; Oh, H. Neural Networks for Pattern Recognition. *Adv Comput.* 1993. Available online: https://books.google.com.hk/books?id=vL-bB7GALAwC&pg=PA165&lpg=PA165&dq=Kothari,+S.C.;+Oh,H.+Neural+Networks+for+Pattern+Recognition.&source=bl&ots=9dkbD_qwsK&sig=ACfU3U16HyCBDuZ2wEYkBNXD5MnuaqQ58Q&hl=zh-TW&sa=X&ved=2ahUKEwiY5rXG0MPlAhXLc94KHWtkAnMQ6AEwAHoECAoQAQ#v=onepage&q=Kothari%2C%20S.C.%3B%20Oh%2C.CH.%20Neural%20Networks%20for%20Pattern%20Recognition.&f=false (accessed on 30 October 2019).
11. Trapero, J.R. Calculation of solar irradiation prediction intervals combining volatility and kernel density estimates. *Energy* **2016**, *114*, 266–274. [CrossRef]
12. Taylor, J.W.; Mcsharry, P.E.; Buizza, R. Wind power density forecasting using ensemble predictions and time series models. *IEEE Trans. Energy Convers.* **2009**, *24*, 775–782. [CrossRef]
13. Khosravi, A.; Nahavandi, S.; Creighton, D.; Atiya, A.F. Lower upper bound estimation method for construction of neural network-based prediction intervals. *IEEE Trans. Neural Netw.* **2011**, *22*, 337–346. [CrossRef] [PubMed]

14. Quan, H.; Srinivasan, D.; Khosravi, A. Short-term load and wind power forecasting using neural network-based prediction intervals. *IEEE Trans. Neural Netw. Learn. Syst.* **2017**, *25*, 303–315. [CrossRef] [PubMed]
15. Wan, C.; Niu, M.; Song, Y.; Xu, Z. Pareto optimal prediction intervals of electricity price. *IEEE Trans. Power Syst.* **2017**, *32*, 817–819. [CrossRef]
16. Shi, Z.; Liang, H.; Dinavahi, V. Wavelet neural network based multiobjective interval prediction for short-term wind speed. *IEEE Access* **2018**, *6*, 63352–63365. [CrossRef]
17. Yadav, A.K.; Chandel, S.S. Solar radiation prediction using artificial neural network techniques: A review. *Renew. Sustain. Energy Rev.* **2014**, *33*, 772–781. [CrossRef]
18. Li, Z.; Liu, X.; Chen, L. Load interval forecasting methods based on an ensemble of Extreme Learning Machines. In Proceedings of the IEEE Power and Energy Society General Meeting, Denver, CO, USA, 26–30 July 2015.
19. Kavousi-Fard, A.; Khosravi, A.; Nahavandi, S. A new fuzzy-based combined prediction interval for wind power forecasting. *IEEE Trans. Power Syst.* **2015**, *31*, 18–26. [CrossRef]
20. Jiang, P.; Li, R.; Li, H. Multi-objective algorithm for the design of prediction intervals for wind power forecasting model. *Appl. Math. Model.* **2019**, *67*, 101–122. [CrossRef]
21. Ak, R.; Li, Y.F.; Vitelli, V.; Zio, E.; Jacintod, C.M.C. NSGA-II-trained neural network approach to the estimation of prediction intervals of scale deposition rate in oil & gas equipment. *Expert Syst. Appl.* **2013**, *40*, 1205–1212.
22. Huang, G.B.; Zhu, Q.Y.; Siew, C.K. Extreme learning machine: Theory and applications. *Neurocomputing* **2006**, *70*, 489–501. [CrossRef]
23. Kasun, L.L.C.; Zhou, H.; Huang, G.; Vong, C. Representational Learning with Extreme Learning Machine for Big Data. *IEEE Intell. Syst.* **2013**, *28*, 31–34.
24. Xiong, L.; Jiankun, S.; Long, W.; Weiping, W.; Wenbing, Z.; Jinsong, W. Short-term wind speed forecasting via stacked extreme learning machine with generalized correntropy. *IEEE Trans. Ind. Inform.* **2018**, *14*, 4963–4971.
25. Eberhart, R.; Kennedy, J. Particle swarm optimization. In Proceedings of the IEEE International Conference on Neural Networks, Perth, Australia, 27 November–1 December 1995; Volume 4, pp. 1942–1948.
26. Storn, R.; Price, K. Differential evolution—A simple and efficient heuristic for global optimization over continuous spaces. *J. Glob. Optim.* **1997**, *11*, 341–359. [CrossRef]
27. Kirkpatrick, S.; Gelatt, C.D.; Vecchi, M.P. Optimization by simulated annealing. *Science* **1983**, *220*, 671–680. [CrossRef] [PubMed]
28. Geem, Z.W.; Kim, J.H.; Loganathan, G.V. A New Heuristic Optimization Algorithm: Harmony Search. *Simulation* **2001**, *76*, 60–68. [CrossRef]

© 2019 by the authors. Licensee MDPI, Basel, Switzerland. This article is an open access article distributed under the terms and conditions of the Creative Commons Attribution (CC BY) license (http://creativecommons.org/licenses/by/4.0/).

Article

A Novel Ensemble Algorithm for Solar Power Forecasting Based on Kernel Density Estimation

Mohamed Lotfi [1,2], Mohammad Javadi [2], Gerardo J. Osório [3], Cláudio Monteiro [1] and João P. S. Catalão [1,2,*]

1. Faculty of Engineering, University of Porto, 4200-465 Porto, Portugal; mohd.f.lotfi@gmail.com (M.L.); cdm@fe.up.pt (C.M.)
2. INESC TEC, 4200-465 Porto, Portugal; msjavadi@gmail.com
3. C-MAST, University of Beira Interior, 6201-001 Covilha, Portugal; gjosilva@gmail.com
* Correspondence: catalao@fe.up.pt

Received: 30 October 2019; Accepted: 29 December 2019; Published: 2 January 2020

Abstract: A novel ensemble algorithm based on kernel density estimation (KDE) is proposed to forecast distributed generation (DG) from renewable energy sources (RES). The proposed method relies solely on publicly available historical input variables (e.g., meteorological forecasts) and the corresponding local output (e.g., recorded power generation). Given a new case (with forecasted meteorological variables), the resulting power generation is forecasted. This is performed by calculating a KDE-based similarity index to determine a set of most similar cases from the historical dataset. Then, the outputs of the most similar cases are used to calculate an ensemble prediction. The method is tested using historical weather forecasts and recorded generation of a PV installation in Portugal. Despite only being given averaged data as input, the algorithm is shown to be capable of predicting uncertainties associated with high frequency weather variations, outperforming deterministic predictions based on solar irradiance forecasts. Moreover, the algorithm is shown to outperform a neural network (NN) in most test cases while being exceptionally faster (32 times). Given that the proposed model only relies on public locally-metered data, it is a convenient tool for DG owners/operators to effectively forecast their expected generation without depending on private/proprietary data or divulging their own.

Keywords: forecasting; ensemble methods; kernel density estimation; smart grids; distributed generation; solar PV

1. Introduction

Accurate prediction of power generation from renewable energy sources (RES) is a challenging task, posing problems for short-term operation of modern power systems [1]. This difficulty is due to the high uncertainties and complexity of both the associated variables and the equipment used for generation and grid connection. On the one hand, generation from RES is a function of multiple meteorological factors (temperature, humidity, wind flow, etc.) which are in and of themselves highly chaotic in nature and difficult to quantify [2,3]. On the other hand, the equipment used is also a source of significant uncertainty with reliability issues and failures commonly occurring in installed power electronics, inverter-side, grid-side, and even the metering apparatus [4]. The combined effect of chaotic input variables and complex energy conversion models render deterministic approaches infeasible for the prediction of distributed generation (DG) from RES. As such, statistical and/or probabilistic models are commonly employed not only to forecast DG but also to predict market behavior in the case of high RES deployment [5,6] which allows for a computationally efficient way of accounting for uncertainties in inputs.

In recent years, there has been increased interest in the use of ensemble methods for power system applications. Ensemble techniques have a decades-long track record in meteorological prediction, proving their potential to effectively predict highly chaotic processes [7].

The main premise of ensemble methods is to overcome both input and model uncertainties by compiling a set (ensemble) of separate predictions into a forecast of most likely outcomes. Each separate prediction is a result of varying input variables within their uncertainty range in addition to the model uncertainty. Therefore, a combination of these separate predictions yields a range of possible outputs representing a confidence/uncertainty region surrounding a most likely scenario.

In Figure 1, the concept of an ensemble forecast is visualized considering the case of DG production from RESs. Various meteorological factors are independent input variables and are associated with a significant level of uncertainty. In addition, the physical energy conversion models of DG units are also associated with a high uncertainty, leading to a significant change in energy generation as a result of small perturbances in the meteorological variables. Ensemble methods combine different scenarios based on both input and model uncertainties and establish a confidence interval around a most likely outcome. One can see that the employment of an ensemble technique involves the (continuously improving) prediction of some variable based on historical data, without knowledge of the physical model relating the inputs with the outputs. This is, in fact, the definition of machine learning (ML), and, as such, ensemble methods are often classified accordingly [8].

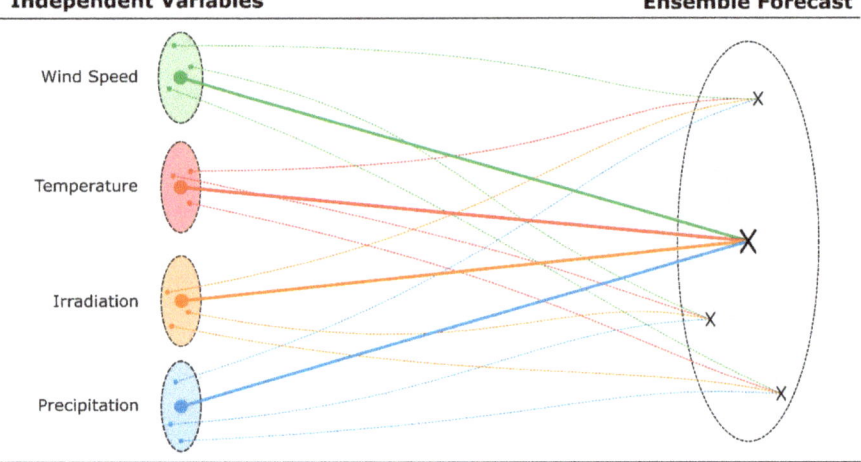

Figure 1. Visualization of an ensemble forecast. Rather than employing a deterministic/point method (thick lines) to obtain the output from input variables, an ensemble of predictions is made from varied input conditions (dashed lines), constructing an uncertainty region and most likely output value(s).

Paper Organization

This manuscript is organized as follows: Section 1 provides the motivation behind this study and an introduction to ensemble forecasting. Section 2 provides a comprehensive state-of-the-art review of recent literature on the topic. Section 3 describes the mathematical formulation of the proposed model. Section 4 presents the case study based on a solar photovoltaic (PV) installation in the center region of Portugal. Section 5 provides the simulation results and a comparison between the proposed method, a deterministic irradiance-based prediction, and a neural network (NN) approach. A discussion of the obtained results is then presented followed by prospects for future work following up on the current study. The conclusions are finally summarized in Section 6. More detailed supplementary data regarding the case study used are provided in Appendix A.

2. State-of-the-Art and Novel Contributions

As mentioned, the use of ensemble methods is gaining popularity with the increased complexity and uncertainty of distributed energy resources (DERs). Before presenting the proposed method, a review of recent works is presented to highlight the state-of-the-art scientific literature on ensemble methods applications to power systems in recent years.

2.1. State-of-the-Art

In Reference [9], different strategies for combining forecasts of solar photovoltaic (PV) generation were presented. In this study, the ensemble prediction was obtained by combining different probabilistic models rather than an ensemble of results of the same model. It used three models (i.e., QKNN, QRF, and QR) and the inputs were historical PV power and weather data.

By testing using the GEFCOM 2014 data, the results showed that the use of an ensemble of various probabilistic forecasts resulted in a significant increase in forecasting accuracy for solar photovoltaic (PV) systems as opposed to the use of individual ones, regardless of the ensemble strategy and/or scenarios considered. In Reference [10], the advantages and disadvantages of applying an ensemble to improve empirical mode decomposition (EMD) techniques were reported, which are mentioned as being commonly applied to wind forecasting. While the ensemble improved EMD models are associated with additional computational burden, they are reported to outperform other techniques, specifically in tackling the challenge of mode mixing. In addition, the authors reported it was significantly more beneficial to apply ensemble decomposition to artificial neural network (ANN) models as compared to using optimization methods to tune the ANN parameters. The previous statements were shown to hold for all time resolutions in wind power forecasting. In Reference [8], a comparison of numerous commonly used ensemble, ANN, and other ML techniques was performed for solar power forecasting. Random Forest (RF), an ensemble method, was found to exhibit the best performance. Two main conclusions were made by the study: (1) a seasonal bias was shown with spring and winter being more challenging to forecast than summer and autumn (keeping in mind that the data were from Norwich, UK) and, more importantly, that (2) a combination of simple algorithms yielded better and more reliable results than any individual algorithm on its own, regardless of its complexity.

In Reference [11], a short-term probabilistic forecasting method was proposed based on a competitive ensemble of different base predictors of PV power. The method was implemented using different probabilistic approaches which were trained as base predictors in order to obtain an ensemble of the predictive distribution with optimal characteristics of accuracy and reliability. In Reference [12], the reliability, robustness, and computational burden of a proposed PV power forecasting model based on the RF method was combined with the extra trees technique on an hourly basis and compared against supervised support vector regression. For a fair and comparative analysis, the models used comparable forecasting data, applicable for forecasting hourly PV power.

A probabilistic PV power forecasting model was proposed in Reference [13] and applied to several French PV plants considering six days of lead time with a resolution of thirty minutes. The proposed model was derived from multiple forecasts considering the national numerical weather predictions and including ensemble forecasts. Then, a free online parameter learning technique generated a weighted combination of the individual PV outputs, and the resulting weights were later sequentially computed before each forecast, using only historical data, with the goal of minimizing the continuous ranked probability score criterion.

An analog ensemble forecasting method for day-ahead regional with hourly resolution was presented in Reference [14]. The proposed model considered publicly available weather forecasts and power measurement data, considering some historical sets of temperature, irradiance, and terrain slopes as well, among others. To process the input data, clustering and blending strategies were used to improve the PV forecasting results which were compared and validated against several numerical models based on weather forecasts.

Photovoltaic power variability was studied in Reference [15], proposing a data-driven ensemble modeling technique to improve the forecasting of PV output. Also, three different models were analyzed within a recursive arithmetic average technique, considering stand-alone forecasting results. To prove the superiority of the proposed model, the comparison was carried out considering a considerable number of different training and testing samples, showing that the ensemble model generally outperforms different stand-alone forecasting models.

A PV forecasting model in Reference [16] used an ANN ensemble scheme based on particle swarm optimization with trained feed-forward neural network. The proposed model was constructed considering five different structures with varying network complexities, in order to improve the forecasting results. Then, the model was combined using trim aggregation after removing the error boundaries. Exogenous data, such as physical specification and environmental, were used as model inputs. Moreover, a clearness index was used to classify days accordingly with their features, considering a yearly basis analysis with a real case study. It was shown that ensemble schemes improve the forecast results in comparison with benchmark models.

In Reference [17], an hourly PV power forecasting model was presented based on clustering and ensemble prediction using the RF method. First, clustering was used to improve the computational burden by selecting the necessary weather variables. Then, the RF method with different parameters was implemented as a component model to find weather regimes making up the ensemble prediction. Finally, weighted computation was carried to analyze the different forecasting weather regimes in order to obtain the final results. Ridge regression was used to determine the weight of each weather variable automatically.

In Reference [18], a hybrid PV forecasting model combined the ML method with the Theta statistical method. Multiple ML components were used: long short-term memory, gate recurrent unit, and unsupervised learning. Structural and data diversity were key to improving the accuracy of the model. Four different approaches were implemented for validation, considering two real case studies. The proposed hybrid model was shown to be superior to traditional ML without statistical components.

In Reference [5], a new ensemble technique was employed to improve probabilistic forecasting of day-ahead price forecasting of the Iberian market. An approach based on kernel density estimation (KDE) was used to "activate" the best set of input variables which minimize the forecasting error. This study is an example of numerous others applying probabilistic and ML techniques for electricity price forecasting which has been increasing exponentially in the past decade as shown by Reference [6]. The latter shows that, while non-existent before 2003, probabilistic methods (or hybrid ones) have quickly gained ground as one of the main approaches used contemporarily for price forecasting [19–21]. The analysis in Reference [22] has shown that, for the case of price forecasting, while combining different forecasts in an ensemble framework does not necessarily always bring about improved accuracy, it does contribute to more reliable forecasting by decreasing the risk associated with an individual method.

Based on the conducted literature review, the following points were noted and were carried forth in the formulation, analysis, and discussion made throughout this paper:

- The use of combinatorial ensemble techniques is shown to significantly improve the accuracy of RES-based DG forecasting in addition to guaranteeing a more reliable and/or robust prediction;
- An ensemble of simple probabilistic/statistical techniques is shown to produce better and more robust DG forecasting than individual complex models;
- KDE has been recently employed to "activate" input sets for probabilistic price forecasting models, showing great success in improving the accuracy. This was only found to be tested on price forecasting, and no studies were found using this methodology for DG forecasting [5].

2.2. Novel Contributions

In this study, we proposed an ensemble algorithm based on the following key points:

1. The objective was to develop an algorithm suitable for predicting DG from RESs. The specific focus of this study was on solar PV; however, the proposed approach is generalizable;
2. Only historical, publicly available data (e.g., meteorological forecasts) and the corresponding local output (i.e., recorded power generation) were given as inputs (i.e., no knowledge of any physical model was known and no dependence on private/proprietary data were needed);
3. The algorithm can run despite inconsistency or loss of data points. Using KDE, the most suitable inputs are "activated" from the historical dataset.

3. Proposed Methodology

Consider an output variable P that has a value which depends on a set of inputs $V := \{v_1, v_2, \ldots N_V\}$ through some unknown model f:

$$P = f(V) = f(v_1, v_2, \ldots N_V) \tag{1}$$

where N_V is the number of independent variables which affect output P. For the purpose of generalization, the inputs V are considered multidimensional, such that:

$$v_1 = \{v_{1,1},\ v_{1,2},\ \ldots v_{1,H_1}\} \tag{2}$$

In this case, H_1 is the number of dimensions of v_1. Now, consider scenario "new" for which we are trying to predict the output P^{new}, given a set of conditions V^{new}:

$$P^{new} = f(V^{new}) = f\left(v_1^{new}, v_2^{new}, \ldots N_V^{new}\right) \tag{3}$$

The goal is to predict the value of P^{new} given only the new conditions V^{new} and a historical set of N_o cases (with no knowledge of f):

$$P^{old,o} = f\left(V^{old,o}\right) = f\left(v_1^{old,o}, v_2^{old,o}, \ldots N_V^{old,o}\right);\ \forall\ o = 1, 2, \ldots, N_o \tag{4}$$

While the model function f is assumed to be chaotic, in this model we assume that the number of independent input variables and their dimensions remain constant and, therefore, the following equations hold:

$$N_V^{new} = N_V^{old,o} = N_V;\ H_i^{new} = H_i^{old,o} = N_H;\ \forall o = 1,2,\ldots, N_o;\ i = 1, 2, \ldots, N_V \tag{5}$$

At this stage, the objective was to select a subset of N_S cases which were most suitable to form an ensemble prediction of P^{new}. To do this, the KDE function similar to Reference [5] was used to calculate a similarity index $s_{old,new}$ between the new case and each of the old cases in the historical dataset. In this case, the most similar N_S cases (with the highest similarity index) can be activated by means of the product of kernel functions of each variable. This is visualized in Figure 2.

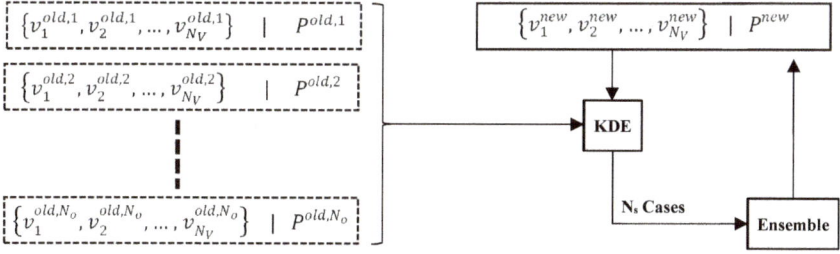

Figure 2. Demonstration of how the proposed Kernel Density Estimation (KDE)-based similarity index is used to extract N_S cases to form an ensemble prediction of the new output value.

The Gaussian kernel functions were used to construct the similarity index KDE, as they are most suitable for cases when little or no knowledge of the model is known.

$$S_{old,new} = \left(\prod_{i}^{N_V} \prod_{j}^{N_H} e^{-\frac{1}{2}\left(\frac{v_{i,h}^{old} - v_{i,h}^{new}}{b_i}\right)^2} \right)^{\frac{1}{N_V N_H}} \quad (6)$$

The bandwidth value b_i can be used to increase or decrease the sampling window (relative to the full range of the historical samples) for each variable in the same manner that KDE works, i.e., the narrower the bandwidth, the higher assumed correlation between variable v and the output P. Therefore, the value of b_i for each variable i can be expressed by means of a tuning coefficient α_i:

$$b_i = \alpha_i \left(\max_o \left(v_i^{old,o} \right) - \min_o \left(v_i^{old,o} \right) \right); \quad \forall \quad o = 1, 2, \ldots, N_o \ ; \ i = 1, 2, \ldots, N_V \quad (7)$$

In this way, this normalized tuning coefficient is varied from 0 (exclusive) to 1 (inclusive), corresponding to a bandwidth value between zero (exclusive) and the maximum range of the historical value of the variable (inclusive):

$$0 < \alpha_i \leq 1 \quad \forall \quad i = 1, 2, \ldots, N_V \quad (8)$$

The similarity index in Equation (6) can be simplified in case all input independent variables are scalars. In this case, N_H is equal to one, and the equation is reduced accordingly:

$$S_{old,new} = \left(\prod_{i}^{N_V} e^{-\frac{1}{2}\left(\frac{v_i^{old} - v_i^{new}}{b_i}\right)^2} \right)^{\frac{1}{N_V}} \quad (9)$$

Given a new case, the similarity index is calculated for all old cases in the historical dataset. We can now construct a sorted array S which has elements that correspond to the index of the old case; this array thus contains the indices of the historical dataset, sorted from most to least similar to the current case based on their calculated similarity index for Equation (9):

$$S = [k_1, k_2 \ldots k_{N_o}] \quad (10)$$

In this case, k_1 is the index o of the historical case with the highest, k_2 to the second highest, etc. Now, the top N_s samples can be selected to perform the ensemble prediction. The simplest prediction is to calculate the mean value of the top N_s P^{old} values:

$$\hat{P}^{new} \approx \frac{\sum_{i=1}^{N_s} P_{k_i}^{old}}{N_s} \quad (11)$$

To obtain a confidence/uncertainty interval around this expected output, percentile ranks can be used by constructing a cumulative distribution function of the top N_s values. By doing so, a confidence interval can be determined as follows:

$$\hat{P}_{lb,x\%}^{new} \leq \hat{P}^{new} \leq \hat{P}_{ub,x\%}^{new} \quad (12)$$

$$\hat{P}_{lb,x\%}^{new} = \rho_{\frac{1}{2}(100-x)\%} \left(\{ P_{k_1}^{old}, P_{k_2}^{old}, \ldots, P_{k_{N_s}}^{old} \} \right) \quad (13)$$

$$\hat{P}_{ub,x\%}^{new} = \rho_{\frac{1}{2}(100+x)\%} \left(\{ P_{k_1}^{old}, P_{k_2}^{old}, \ldots, P_{k_{N_s}}^{old} \} \right) \quad (14)$$

What Equation (12) means is that for a confidence of $x\%$, $\hat{P}^{new}$ lies between the lower and upper bounds equal to $\hat{P}_{lb,x\%}^{new}$ and $\hat{P}_{ub,x\%}^{new}$, respectively; which are calculated, as per Equations

(13) and (14), by means of the percentile $\rho_{\frac{1}{2}(100-x)\%}$ and $\rho_{\frac{1}{2}(100+x)\%}$ of the top N_S values $\left(\left\{P_{k_1}^{old}, P_{k_2}^{old}, \ldots, P_{k_{N_S}}^{old}\right\}\right)$, respectively.

It must be noted that there are clearly more complex means of calculating $\hat{P}^{new}$ and the confidence bounds. However, the main focus of this study was to highlight the use of the similarity index to extract the set S, and the choice of the simplest ensemble prediction afterwards was intentional to demonstrate the power of such a selection algorithm even with the most basic ensemble applied.

4. Case Study and Validation

4.1. PV Installation in Portugal

In order to test and validate the proposed algorithm, a real case study was used based on solar PV installations located in the vicinity of the city of Coimbra in the center region ("*Região do Centro*") of Portugal as shown in Figure 3. The technical specifications of the plant are listed in Table 1. Historical forecasts and measurements are available for the same installation for a full year from 15 March 2015 to 15 March 2016 as detailed in Table 2. Annual plots of all variables are provided in Appendix A1.

In this case, the historical weather forecasts were the input variables (V) and are publicly provided by the Global Forecasting System (GFS) model with a 22 km resolution. The GFS's data are available for any region of the world and is publicly available online [23]. The forecasts are made at 18:00 (UTC time) of each day for the day-ahead with a 3 h resolution (average of each 3 h interval of the day: 0:00, 3:00, 6:00, ..., 21:00). The provided forecasts are for wind speed, temperature, solar irradiance, precipitation, and humidity.

The output AC power of the inverter was recorded for the same year. A 20 kW SMA Sunny Tripower inverter was installed with 2 maximum power point trackers (MPPTs) installed (4 strings per inverter). The logging frequency of the AC power output was approximately 5 min. For this study, the recorded AC power was synchronized with the forecasts by applying a 3 h average (averaging can be seen in Figure A6). It is important to stress that the proposed prediction method was only given the averaged output power as input. However, high-resolution data were used for validation to test if uncertainties associated with high frequencies were captured.

4.2. Numerical Irradiance-Based Forecast

Given that the GFS data and the output power were synchronized, and since MPPTs were installed with the inverters, one can use the following equation to predict the maximum possible power output from the current installation for each data point.

$$P_t \approx P_t^{irr} = \eta_{avg} N_p A_p R_t^{wf} \tag{15}$$

where P_t^{irr} is the predicted power output at time t calculated numerically from the irradiance forecast, η_{avg} is the overall average energy conversion efficiency of the PV plant (accounting for the PV conversion and inverter efficiency), N_p and A_p are the number of panels and the area of each panel (in m^2), respectively, and R_t^{wf} is the direct incident solar irradiance (W/m^2) obtained from the weather forecast for time t.

Table 1. Technical specifications of the solar photovoltaic (PV) plant used as a case study.

Parameter	Value	Units
Number of Panels (300 kWp each)	53	-
Panel Area (each)	1.713	m^2
Total Installed Capacity	18	kWp
Inverter Capacity	20	kW
Nominal DC Voltage	600	V
Overall Efficiency	20	%

Table 2. Details of variables in the historical dataset provided (from 15 March 2015 to 15 March 2016).

Historical Variable	Data Source	Spatial Resolution	Temporal Resolution	Units
Wind Speed	Meteorological Forecast	22 km	3 h	m/s
Temperature	Meteorological Forecast	22 km	3 h	°C
Solar Irradiance	Meteorological Forecast	22 km	3 h	W/m^2
Precipitation	Meteorological Forecast	22 km	3 h	mm
Humidity	Meteorological Forecast	22 km	3 h	%
Inverter AC Power (Output)	Real Measurement	-	~5 min	kW

Figure 3. Region in the center of Portugal used as a case study. The PV installations used in the current analysis were located within a 10 km radius of the city of Coimbra (40°12′ N, 8°25′ W).

4.3. Seasonal Test Weeks

Also, in order to check for seasonal effects and/or bias, four test weeks were extracted from the annual data corresponding to all four seasons. The annual measured output power, annual predicted maximum output (based on irradiance estimation in Equation (15)), and detailed plots thereof for all four representative weeks are shown in Figure 4.

By inspecting the plots shown in Figure 4, particularly comparing the maximum theoretical output based on irradiance and recorded power, two important observations are worthy of noting:

- During the summer, the maximum power output prediction based on Equation (15) was greater than the recorded value. This is what one would expect, and the operating efficiency and/or reliability of the installation would seldom reach the maximum theoretical power output;
- During the winter, the prediction based only on solar irradiation failed to predict any value of output power (one can see that the predicted values were zeros throughout the winter and also by looking at the plot of the winter week). This is due to the fact that the meteorological forecasts provided by GFS are averaged over large temporal and spatial resolutions. As such, the forecasted irradiance would dissipate during winter weather conditions.

As such, it is clear that relying solely on the irradiance models, is insufficient to make any prediction of the expected power output of the solar PV installations.

Therefore, the objective of this case study was to check if the proposed method, taking into consideration GFS data as input variables and the recorded (and synchronized) AC power output of the plant, would be capable of accurately forecasting the power output under different meteorological conditions.

The GFS meteorological data are plotted for the entire year in Figures A1–A6 in the Appendix A. Zoomed-in plots are also provided for each test week in order to show the seasonal differences and highlight some visible correlation between the weather conditions and the recorded AC power output. The plots of spring, summer, autumn, and winter are shown in Figures 5–8, respectively.

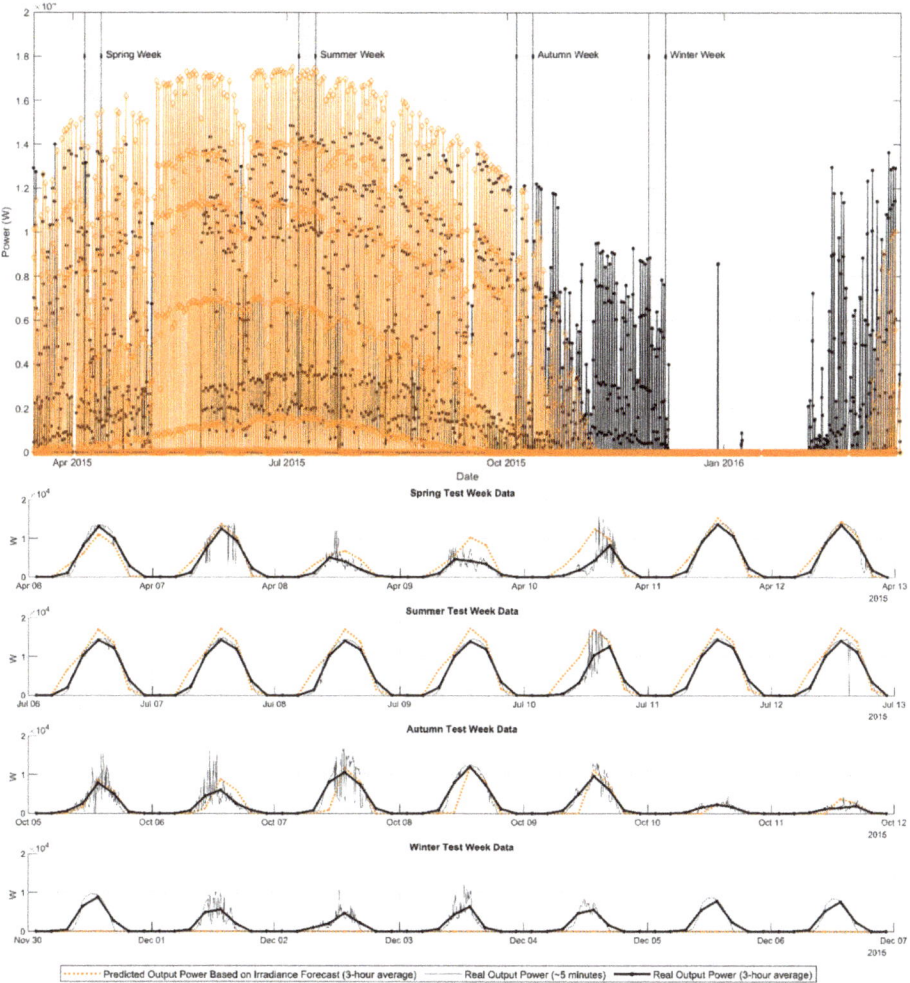

Figure 4. Annual plot of recorded AC power output, annual plot of maximum theoretical power output based on solar irradiance estimation and average efficiencies, and four test weeks representing all four seasons (**top**); and for each test week, zoomed-in plots of recorded AC power output (un-averaged), 3 h averaged recorded AC power output (synchronized with GFS data), and maximum theoretical power output based on solar irradiance estimation and average efficiencies (**bottom**).

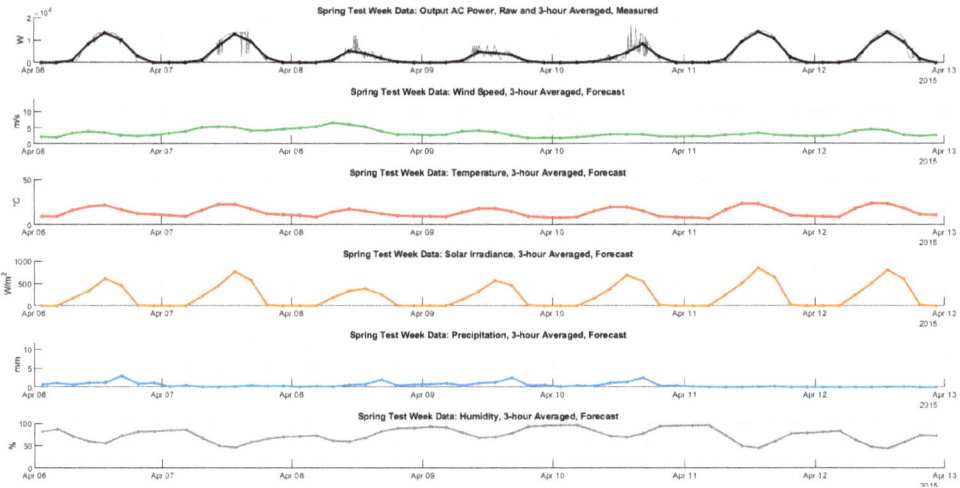

Figure 5. Plots of recorded output power and Global Forecast System (GFS) meteorological data for the spring test week.

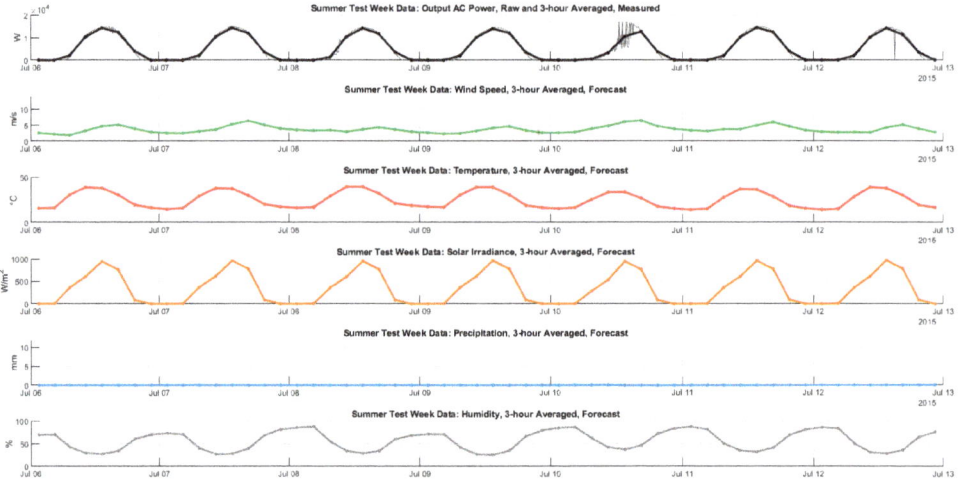

Figure 6. Plots of recorded output power and GFS meteorological data for the summer test week.

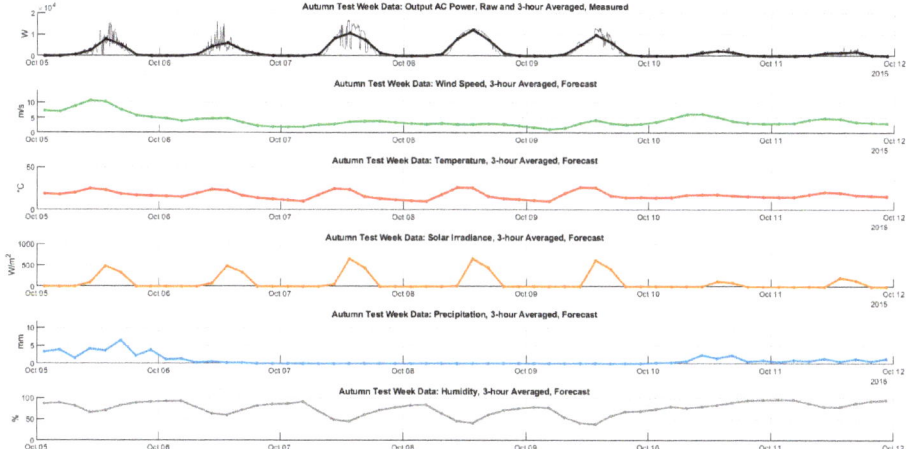

Figure 7. Plots of recorded output power and GFS meteorological data for the autumn test week.

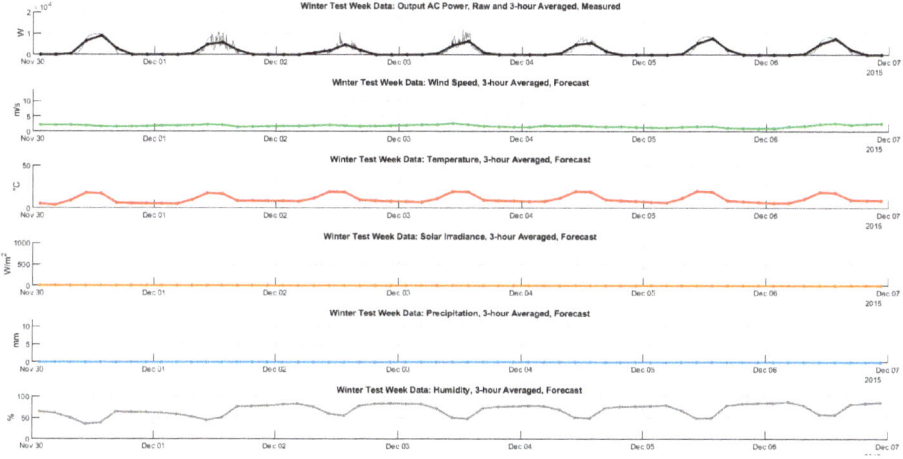

Figure 8. Plots of recorded output power and GFS meteorological data for the winter test week.

4.4. Implementation and Validation

To test the proposed algorithm in Section 2, the power output for each of the four test weeks was forecasted, only taking as input variables the meteorological forecasts provided by GFS. The hour of the day and day of the year were appended to the array of input variables in order to give the potential of favoring closer times/dates. The input variable array for this case was as follows:

$$V = \{v_1, v_2, v_3, v_4, v_5, v_6, v_7\} \tag{16}$$

The description of each variable and the choice of the bandwidth tuning coefficients (Equations (7) and (8)) are provided in Table 3. As explained in Section 2, the smaller the value of α, the higher the assumed correlation between the output variable and its corresponding input variable. The values used in this study were assumed based on the well-established physical relationships between each meteorological variable and the target one (PV output power). For instance, solar irradiance was associated with the most dependence and thus a value of 0.1 was chosen, etc. This can heuristically be set based on visual inspection of Figures 5–8.

Table 3. Description of input variables for the historical dataset and value chosen for bandwidth coefficient for the KDE-based similarity index calculator.

Bandwidth Coefficient	Value
a_{v1} (hour of the day)	0.4
a_{v2} (day of the year)	1.0
a_{v3} (wind speed forecast)	0.8
a_{v4} (temperature forecast)	0.5
a_{v5} (solar irradiance forecast)	0.1
a_{v6} (precipitation forecast)	0.8
a_{v7} (humidity forecast)	0.5

In order to investigate the performance of the proposed algorithm, the results obtained for the four test weeks are against the numerical irradiance-based forecast based on Equation (15), and an ANN (trained using the same data). A feed-forward ANN was used with 1 hidden layer and 10 neurons. The performance of all three methods was compared in terms of computational time and accuracy. Since the ANN was trained using the Levenberg–Marquardt algorithm [24], its results and computational time both varied in every run due to the random data division and training process employed. Therefore, to evaluate the results in a fair manner, the ANN was run a sufficiently large number of times (10,000 runs), and the average runtime and forecast results were used for comparison.

To quantify the forecast error, three criteria were used: the mean absolute error (MAE), root mean square deviation (RMSD), and the normalized root mean square deviation (NRMSD). The MAE provides a simple overall measurement of the mean error between forecasted ($\hat{P}$) and real (P) values:

$$MAE = \frac{\sum_{t=1}^{N_t}|\hat{P}_t - P_t|}{N_T} \quad (17)$$

where the subscript t corresponds to the value at time step t and N_t is the total number of time steps. The RMSD is based on the on the quadratic mean:

$$RMSD = \sqrt{\frac{\sum_{t=1}^{N_t}(\hat{P}_t - P_t)^2}{N_T}} \quad (18)$$

Both the MAE and RMSD provide a scale-dependent measure of the deviation between the forecasted and real values. The NRMSD provides a normalized measure as a percentage which is sometimes more favorable when comparing different models.

$$NRMSD = \frac{RMSD}{(P_{max} - P_{min})} \cdot 100\% \quad (19)$$

P_{max} and P_{min} are the maximum and minimum values of the real data, respectively. As such, the NRMSD provides a scale-independent measure. The MAE, RMSD, NRMSD, and computational time are all used to assess the performance of the different approaches for all four test weeks.

The proposed algorithm was developed as original code by the authors using the MATLAB R2019b environment on a standard laptop computer with the following specifications: Intel Core i7-8550U CPU @ 1.80 GHz, 16.0 GB RAM, Windows 10 64 bit operating system. The neural network used for validation was based on the MATLAB 2019b Statistics and Machine Learning Toolbox [24].

5. Results and Discussion

5.1. Results of the Proposed Ensemble Algorithm

The results of the proposed ensemble algorithm are shown in Figure 9. The predicted value was plotted, along with confidence intervals of 68%, 95%, and 99.7%. The following points are noted:

- The proposed ensemble algorithm successfully managed to forecast the wind power output, relying only on the historical GFS meteorological data, for all four tests weeks of all seasons;
- The power production in cases when the deterministic model based on irradiance was inadequate (i.e., winter season) was successfully predicted;
- Despite only being provided averaged data, the confidence intervals successfully managed to cover high-frequency fluctuations during most days;
- The confidence interval grows and shrinks in response to such fluctuations even within the same day (e.g., Summer week, day 5);
- The forecasted mostly underestimated the power output than. This is favorable to overestimation particularly from the point of view of operators of DG installations.

5.2. Comparison and Validation

A comparison between the forecast obtained and that of an irradiance-based numerical model (Equation (15)) and an ANN was used to validate the proposed method. As elaborated in the previous section, the same data were used to train the ANN. Since a random data division and training method was employed (which aimed to minimize the computational time of the ANN), the average of a sufficiently large number of runs of the ANN (i.e., 10,000 runs) was used for a fair comparison.

The comparison was made considering the MAE, RMSD, and NRMSD error criteria for each of the test weeks and is shown in Table 4. The different forecasts are visualized in the plots shown in Figure 10. The computational time to forecast all four weeks by the proposed method and the ANN (average of 10,000 runs in each case) is shown in Table 5.

By comparing the results of the different models, the following points can be verified:

- According to all error criteria used, the proposed method outperformed the irradiance-based prediction for all seasons. It outperformed the ANN in all seasons except winter;
- Both the ANN and the proposed method managed to provide a reasonably accurate prediction of the output power in the winter, where a numerical irradiance-based model completely fails;
- Despite the ANN being capable of providing a better average error for the winter, the capability of the proposed method to capture high-frequency fluctuations in its confidence intervals provides an advantage over the ANN;
- The proposed method was extraordinarily fast in terms of computational time, being 32 times faster than the ANN while outperforming the ANN in the majority of situations.

5.3. Prospects for Future Work

After testing the proposed method, confirming its validity, and taking note of its superior performance particularly in terms of providing a highly computationally efficient forecast, the following recommendations are provided for future work following on this study:

- The effect of using additional meteorological variables (e.g., absolute and relative atmospheric pressure) should be investigated in terms of the forecast accuracy and computational burden;
- Optimal tuning of the bandwidth coefficients should be studied. This can be performed in a pre-processing stage (e.g., with correlation analysis) or using a reinforcement learning-based design in which the values are self-tuned every time the code is run. In the latter, using an optimization method to determine the optimal values may be an option for a hybrid structure;
- Due to the fact of its high computational efficiency and its reliance only on publicly available historical weather forecasts, the proposed method seems to have great potential to be applied to forecast RES-based DG. As such, follow-up work should test the proposed method on other RES technologies such as wind power.

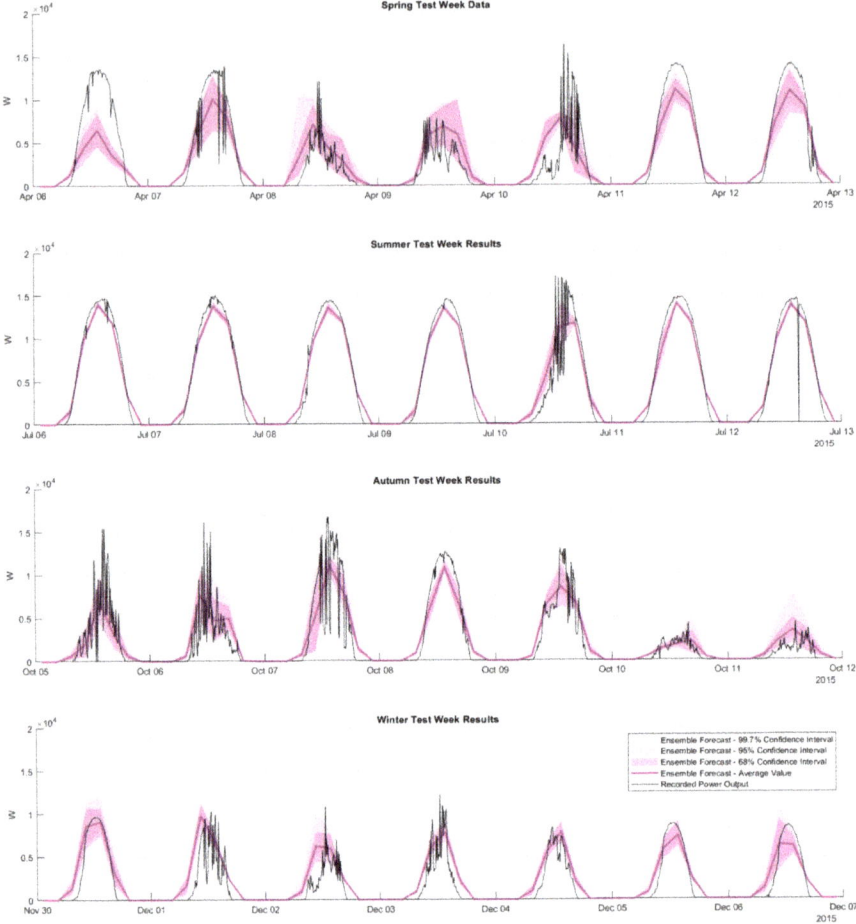

Figure 9. Results of the proposed algorithm for all four seasons, showing real output power (un-averaged) and predicted output power. Confidence intervals of 68%, 95%, and 99.7% are highlighted.

Table 4. Comparison of the MAE, RMSD, and NRMSD error criteria for the results obtained for each of the test weeks from the irradiance forecast, ANN, and the proposed method.

Criterion	Method	Winter	Spring	Summer	Autumn
	Irradiance Forecast	34.6	15.7	17.4	15.3
MAE (kW)	Neural Network	10.7	15.5	8.4	7.9
	Proposed Method	12.6	14.0	3.6	7.7
	Irradiance Forecast	3.062	2.138	2.508	1.857
RMSD (kW)	Neural Network	0.949	2.114	1.203	0.951
	Proposed Method	1.115	1.914	0.523	0.928
	Irradiance Forecast	34.6	15.7	17.4	15.3
NRMSD (%)	Neural Network	10.7	15.5	8.4	7.9
	Proposed Method	12.6	14.0	3.6	7.7

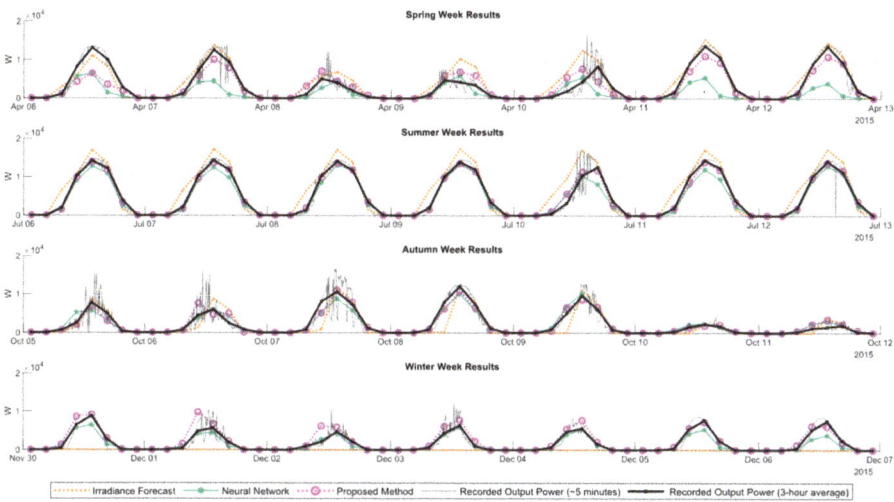

Figure 10. Comparison of results obtained by irradiance forecast estimate, ANN, and the proposed method for all four seasons.

Table 5. Comparison of the computational time between the proposed method and the ANN.

	Computational Time to Forecast all Four Weeks (Average of 10,000 runs)
Neural Network	1.46 s
Proposed Method	0.045 s

6. Conclusions

In this study, a novel ensemble algorithm based on kernel density estimation (KDE) was proposed to forecast RES-based DG, particularly PV power. The proposed method relies solely on publicly available historical series of independent input variables (i.e., historical meteorological forecasts) and the corresponding local output (i.e., recorded power generation). Given a new case (with forecasted meteorological variables), the resulting power generation was forecasted. For the new case to be forecasted, a KDE-based similarity index was used to determine a set of most similar cases from the historical dataset. Then, the corresponding outputs of the most similar cases were used to calculate an ensemble prediction for the forecasted power generation. The proposed method was tested by considering meteorological and recorded power generation from a PV installation around the city of Coimbra, in the center region of Portugal. Despite only being given averaged data as inputs, the developed algorithm was capable of predicting uncertainties associated with high frequency variations in weather conditions, outperforming deterministic prediction based on solar irradiance forecasts. The proposed method outperformed an ANN in most cases while being exceptionally faster (32 times more than the computational time). Given its exceptional computational efficiency and its reliance solely on public data (weather forecasts) and local metered data (power generation), it is a convenient tool for use by owners or operators of solar power installations to effectively forecast their expected generation without depending on private/proprietary data or divulging their own.

Author Contributions: Writing, M.L. and G.J.O.; visualization, M.L., M.J., and G.J.O.; conceptualization, C.M. and M.L.; methodology, M.L. and C.M.; validation, M.L., C.M., G.J.O., M.J., and J.P.S.C.; supervision, C.M. and J.P.S.C. All authors have read and agreed to the published version of the manuscript.

Funding: M.L. would like to acknowledge the support of the MIT Portugal Program (in Sustainable Energy Systems) by Portuguese funds through FCT, under grant PD/BD/142810/2018. M.S. Javadi and J.P.S. Catalão acknowledge the support of the FEDER funds through COMPETE 2020 and by the Portuguese funds through FCT, under POCI-01-0145-FEDER-029803 (02/SAICT/2017).

Conflicts of Interest: The authors declare no conflict of interest.

Appendix A Annual Data

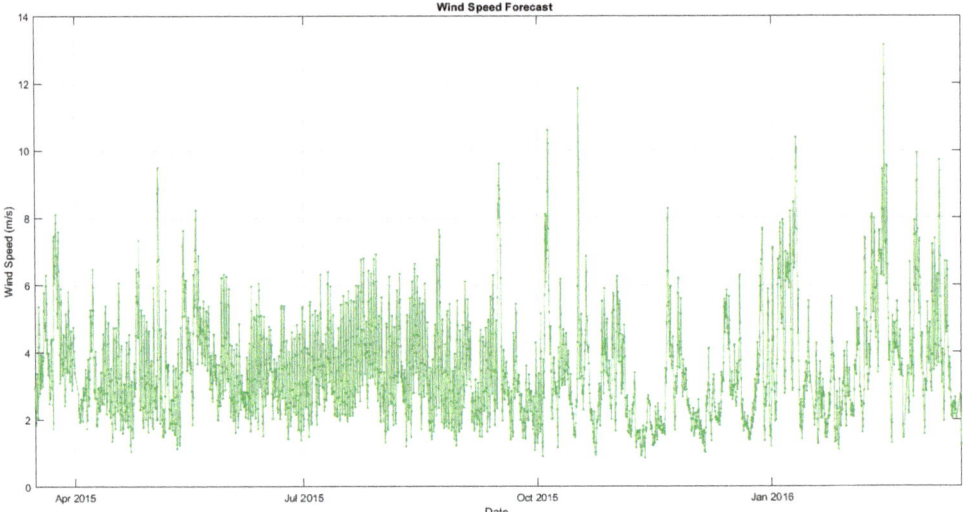

Figure A1. Annual wind speed data for the case study provided by GFS.

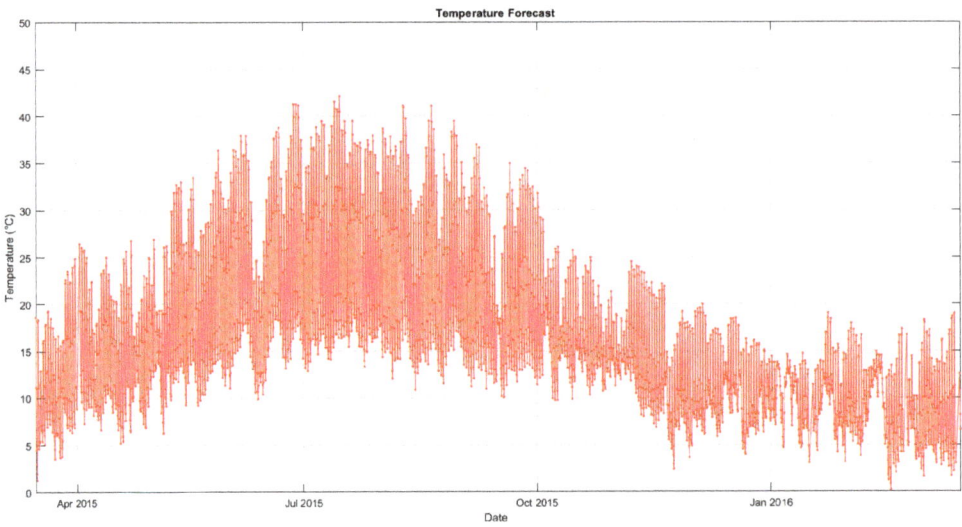

Figure A2. Annual temperature data for the case study provided by GFS.

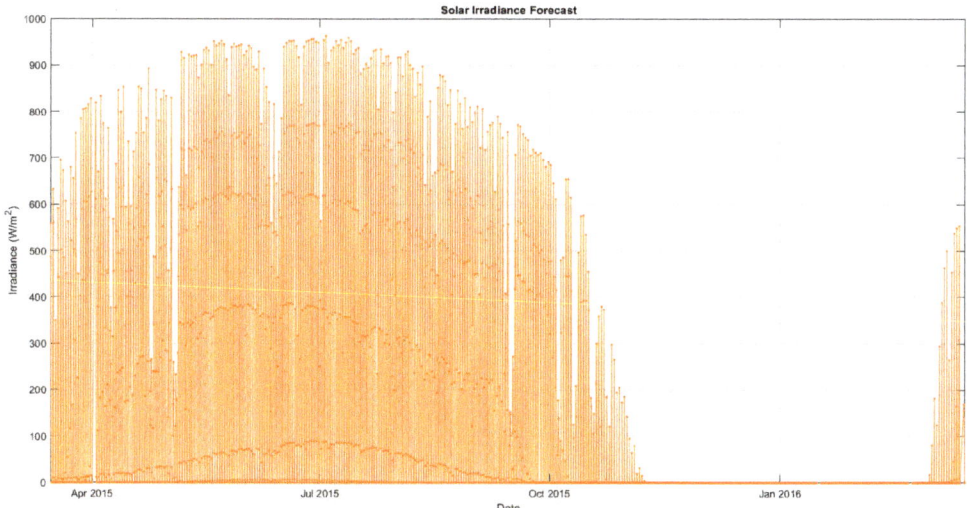

Figure A3. Annual solar irradiance data for the case study provided by GFS.

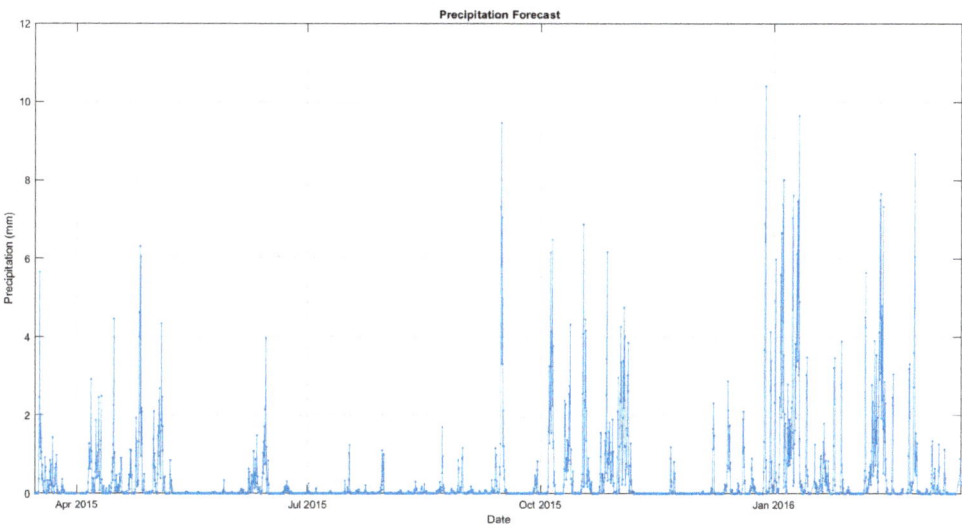

Figure A4. Annual precipitation data for the case study provided by GFS.

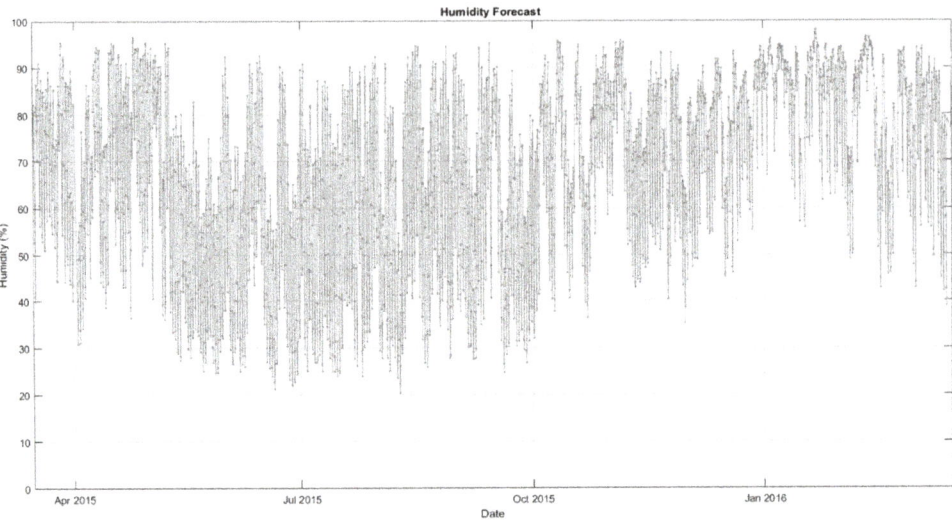

Figure A5. Annual humidity data for the case study provided by GFS.

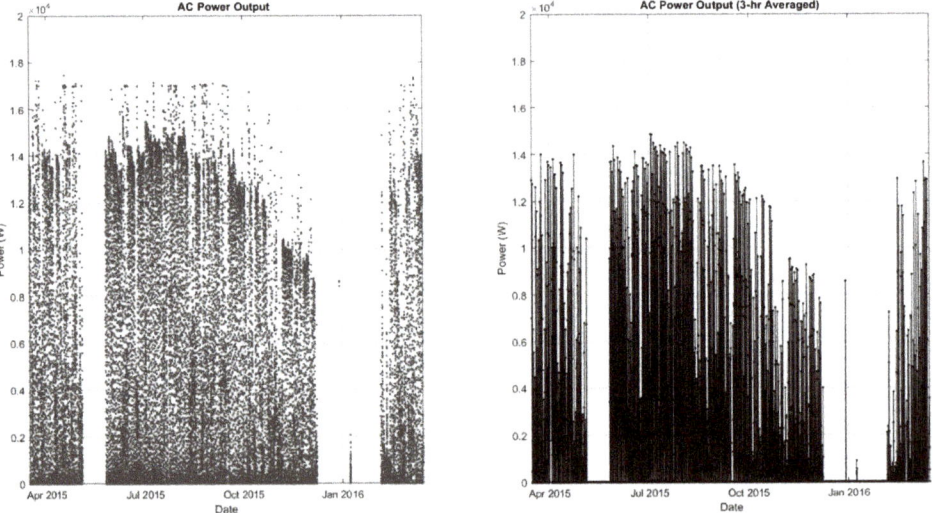

Figure A6. Annual recorded AC output power data for the case study: recorded (**left**) and averaged for GFS synchronization (**right**).

References

1. Kotsalos, K.; Miranda, I.; Silva, N.; Leite, H. A Horizon Optimization Control Framework for the Coordinated Operation of Multiple Distributed Energy Resources in Low Voltage Distribution Networks. *Energies* **2019**, *12*, 1182. [CrossRef]
2. Dev, S.; Alskaif, T.; Hossari, M.; Godina, R.; Louwen, A.; Van Sark, W. Solar Irradiance Forecasting Using Triple Exponential Smoothing. In *2018 International Conference on Smart Energy Systems and Technologies, SEST 2018-Proceedings*; Institute of Electrical and Electronics Engineers Inc.: Piscataway, NJ, USA, 2018. [CrossRef]
3. Gough, M.; Lotfi, M.; Castro, R.; Madhlopa, A.; Khan, A.; Catalão, J.P.S. Urban Wind Resource Assessment: A Case Study on Cape Town. *Energies* **2019**, *12*, 1479. [CrossRef]

4. Formica, T.J.; Khan, H.A.; Pecht, M.G. The Effect of Inverter Failures on the Return on Investment of Solar Photovoltaic Systems. *IEEE Access* **2017**, *5*, 21336–21343. [CrossRef]
5. Monteiro, C.; Ramirez-Rosado, I.J.; Fernandez-Jimenez, L.A.; Ribeiro, M. New Probabilistic Price Forecasting Models: Application to the Iberian Electricity Market. *Int. J. Electr. Power Energy Syst.* **2018**, *103*, 483–496. [CrossRef]
6. Nowotarski, J.; Weron, R. Recent Advances in Electricity Price Forecasting: A Review of Probabilistic Forecasting. *Renew. Sustain. Energy Rev.* **2018**, *81*, 1548–1568. [CrossRef]
7. Palmer, T. The ECMWF Ensemble Prediction System: Looking Back (More than) 25 Years and Projecting Forward 25 Years. *Q. J. R. Meteorol. Soc.* 2018. [CrossRef]
8. Su, D.; Batzelis, E.; Pal, B. Machine Learning Algorithms in Forecasting of Photovoltaic Power Generation. In Proceedings of the 2019 International Conference on Smart Energy Systems and Technologies (SEST), Porto, Portugal, 9–11 September 2019. [CrossRef]
9. Bracale, A.; Carpinelli, G.; De Falco, P. Developing and Comparing Different Strategies for Combining Probabilistic Photovoltaic Power Forecasts in an Ensemble Method. *Energies* **2019**, *12*, 11. [CrossRef]
10. Qian, Z.; Pei, Y.; Zareipour, H.; Chen, N. A Review and Discussion of Decomposition-Based Hybrid Models for Wind Energy Forecasting Applications. *Appl. Energy.* **2019**, *235*, 939–953. [CrossRef]
11. Bracale, A.; Carpinelli, G.; De Falco, P. A Probabilistic Competitive Ensemble Method for Short-Term Photovoltaic Power Forecasting. *IEEE Trans. Sustain. Energy* **2017**, *8*, 551–560. [CrossRef]
12. Ahmad, M.W.; Mourshed, M.; Rezgui, Y. Tree-Based Ensemble Methods for Predicting PV Power Generation and Their Comparison with Support Vector Regression. *Energy* **2018**, *164*, 465–474. [CrossRef]
13. Thorey, J.; Chaussin, C.; Mallet, V. Ensemble Forecast of Photovoltaic Power with Online CRPS Learning. *Int. J. Forecast.* **2018**, *34*, 762–773. [CrossRef]
14. Zhang, X.; Li, Y.; Lu, S.; Hamann, H.F.; Hodge, B.M.; Lehman, B. A Solar Time Based Analog Ensemble Method for Regional Solar Power Forecasting. *IEEE Trans. Sustain. Energy* **2019**, *10*, 268–279. [CrossRef]
15. Liu, L.; Zhan, M.; Bai, Y. A Recursive Ensemble Model for Forecasting the Power Output of Photovoltaic Systems. *Sol. Energy* **2019**, *189*, 291–298. [CrossRef]
16. Raza, M.Q.; Nadarajah, M.; Li, J.; Lee, K.Y.; Gooi, H.B. An Ensemble Framework For Day-Ahead Forecast of PV Output in Smart Grids. *IEEE Trans. Ind. Inform.* **2018**, *15*, 4624–4634. [CrossRef]
17. Pan, C.; Tan, J. Day-Ahead Hourly Forecasting of Solar Generation Based on Cluster Analysis and Ensemble Model. *IEEE Access* **2019**, *7*, 112921–112930. [CrossRef]
18. AlKandari, M.; Ahmad, I. Solar Power Generation Forecasting Using Ensemble Approach Based on Deep Learning and Statistical Methods. *Appl. Comput. Inform.* 2019. [CrossRef]
19. Osório, G.J.; Matias, J.C.O.; Catalão, J.P.S. Electricity Prices Forecasting by a Hybrid Evolutionary-Adaptive Methodology. *Energy Convers. Manag.* **2014**, *80*, 363–373. [CrossRef]
20. Catalao, J.P.S.; Pousinho, H.M.I.; Mendes, V.M.F. Hybrid Wavelet-PSO-ANFIS Approach for Short-Term Electricity Prices Forecasting. *IEEE Trans. Power Syst.* **2011**, *26*, 137–144. [CrossRef]
21. Osório, G.; Lotfi, M.; Shafie-khah, M.; Campos, V.; Catalão, J.; Osório, G.J.; Lotfi, M.; Shafie-khah, M.; Campos, V.M.A.; Catalão, J.P.S. Hybrid Forecasting Model for Short-Term Electricity Market Prices with Renewable Integration. *Sustainability* **2018**, *11*, 57. [CrossRef]
22. Nowotarski, J.; Weron, R. To Combine or Not to Combine? Recent Trends in Electricity Price Forecasting. In *HSC Research Report*; Hugo Steinhaus Center, Wroclaw University of Technology: Wrocław, Poland, 2016.
23. Global Forecast System (GFS) | National Centers for Environmental Information (NCEI) formerly known as National Climatic Data Center (NCDC). Available online: https://www.ncdc.noaa.gov/data-access/model-data/model-datasets/global-forcast-system-gfs (accessed on 14 December 2019).
24. The Mathworks Inc. *Statistics and Machine Learning Toolbox User's Guide R2019*; The Mathworks Inc.: Natick, MA, USA, 2019.

© 2020 by the authors. Licensee MDPI, Basel, Switzerland. This article is an open access article distributed under the terms and conditions of the Creative Commons Attribution (CC BY) license (http://creativecommons.org/licenses/by/4.0/).

Article

Characterising Seasonality of Solar Radiation and Solar Farm Output

John Boland

Centre for Industrial and Applied Mathematics, University of South Australia, Adelaide, SA 5001, Australia; john.boland@unisa.edu.au

Received: 11 November 2019; Accepted: 11 January 2020; Published: 18 January 2020

Abstract: With the recent rapid increase in the use of roof top photovoltaic solar systems worldwide, and also, more recently, the dramatic escalation in building grid connected solar farms, especially in Australia, the need for more accurate methods of very short-term forecasting has become a focus of research. The International Energy Agency Tasks 46 and 16 have brought together groups of experts to further this research. In Australia, the Australian Renewable Energy Agency is funding consortia to improve the five minute forecasting of solar farm output, as this is the time scale of the electricity market. The first step in forecasting of either solar radiation or output from solar farms requires the representation of the inherent seasonality. One can characterise the seasonality in climate variables by using either a multiplicative or additive modelling approach. The multiplicative approach with respect to solar radiation can be done by calculating the clearness index, or alternatively estimating the clear sky index. The clearness index is defined as the division of the global solar radiation by the extraterrestrial radiation, a quantity determined only via astronomical formulae. To form the clear sky index one divides the global radiation by a clear sky model. For additive de-seasoning, one subtracts some form of a mean function from the solar radiation. That function could be simply the long term average at the time steps involved, or more formally the addition of terms involving a basis of the function space. An appropriate way to perform this operation is by using a Fourier series set of basis functions. This article will show that for various reasons the additive approach is superior. Also, the differences between the representation for solar energy versus solar farm output will be demonstrated. Finally, there is a short description of the subsequent steps in short-term forecasting.

Keywords: solar energy; solar farm; clearness index; clear sky index; Fourier series; forecasting

---

1. Introduction

This study is an extension of a paper presented at the 21st International Congress on Modelling and Simulation, Gold Coast, Australia [1]. Further justification of the argument for selecting the additive model for representing the seasonality of solar radiation has been added. Additionally, the discussion has been extended to include the differences for dealing with the seasonality of solar farm output as compared to solar radiation per se.

The literature includes a wide range of methods for forecasting solar radiation on different time scales. Two papers in particular [2,3] contain comprehensive reviews of recent articles in this area. The approaches range from use of Artificial Neural Networks (ANN) using solar irradiation, rather than some transformed variable [4], to several methods where the first step is some type of seasonal adjustment. This can take the form of multiplicative de-seasoning such as using clearness index or a clear sky model, or additive de-seasoning using Fourier series or wavelets. Before looking at these various methods of seasonal adjustment, let us examine the range of forecasting tools apart from that process of the modelling. Forecasting tools cover a broad range from ANN ([4–6] and several other references) to Adaptive Autoregressive [7] to Exponential Smoothing [8]. Several studies make use of what might be called hybrid models, like wavelets plus ANN [9–11], and the Coupled Autoregressive and Dynamical Systems (CARDS) model of the present author and colleagues [12]. This gives a flavour of the wide range of possible methods used for short-term forecasting of solar radiation. These all involve some form of mathematical or statistical approach, but there are also ways of utilising sky cameras, cloud motion vectors, satellite imaging, and so forth.

Seasonality

The methods above, regarding dealing with the inherent seasonality in sub-daily solar radiation series, have to be examined in detail, as characterising the seasonality is the first step in forecasting on hourly and sub-hourly time scales. As mentioned, one approach has been to use multiplicative de-seasoning in the form of dividing the solar radiation series by, in some cases, the extraterrestrial radiation over the site in question for the same time to produce the clearness index [13–16]. Alternatively, numerous articles deal with dividing the solar radiation by some clear sky model to create a clear sky index [5,7].

It is useful to examine whether a multiplicative model is used for describing the seasonality of solar radiation by necessity or for some historical reason. The most usual application of the multiplicative model is for economic series. This is because most seasonal economic series display seasonal variation that increases with the level of the series. For example, in time series that describe tourist arrivals [17], there are more arrivals in particular seasons, but also there can be more variability in those seasons as well. Often practitioners model the seasonality by first taking logarithms of the data in order to stabilise the variance. This approach has been also done with solar data by at least one researcher [18]. This method of using a logarithmic transform coupled with multiplicative de-seasoning for solar data might be a possible method since there is more pronounced variability in the summer months when there is a higher level in the series as well. It will be shown below that using an additive Fourier series representation of the seasonality is a very effective way of dealing with this phenomenon. Apart from this, one could make a case for the use of the clearness index rather than a clear sky model for multiplicative de-seasoning, even though [19] discussed the use of both methods and decided to use the clear sky index. One reason is that Ineichen [20] feels it is necessary to examine the relative efficacy of numerous clear sky models, whereas the clearness index has a well defined formulation. Inman et al. [2] present a poignant discussion on the viability of using clearness index versus clear sky index. The clear sky model requires input of local values of atmospheric variables such as ozone content, water vapour and turbidity. Alternatively, the extraterrestrial radiation only requires inputs such as latitude, time of year, and such like. It does not require data to be measured and input, and thus is not subject to atmospheric fluctuations.

There are many reasons why an additive model to describe the seasonality is more appropriate than a multiplicative one. In particular, a Fourier series approach displays a number of benefits. At any time throughout the year, the Fourier series representation gives the expected value of the variable in question. One could describe this as representing the climate for the location. The difference between the Fourier series model and the data at a particular time can be thought of as the influence of the weather. This could be for solar radiation as is being discussed here, or alternatively ambient temperature, electricity load, or other variables displaying similar seasonal characteristics. This is one of the valuable attributes referred to by Skeiker [21] when he talks of the physical meaning inherent in this representation, which other methods do not necessarily display. Some other researchers, for example, Dong et al. [8], discuss the important sub-diurnal cycles in solar radiation time series, but do not explain their presence from a physical point of view. As will be seen in later sections, the Fourier series approach lends itself very well to an exploration of the physical nature of these cycles. From a statistical viewpoint, simple formulae allow one to calculate the amount of variance of the original data explained by the Fourier series representation for each of the frequencies involved. Arguing against the use of the Fourier series approach, the comments put forward in [2] with respect to clear sky models might also apply here in that for estimating the Fourier coefficients one needs data for some particular period, some years for daily or hourly data down to some months for minute data. However, if ground station data is not available, there is data available that is inferred from satellite images. One could argue that there is no data from satellite models for the minute time scale, but the inherent smoothing provided by the Fourier model at a half hour time scale for instance infers values at lower time scales.

2. Fourier Series Representation

The present author [22] described the physical nature of the significant frequencies that are inherent in the solar radiation data. The yearly and daily cycles are intuitively obvious. The necessity of including the twice daily cycle, also identified by [8] is less obvious. It could represent the fact that as well as night being different from day, morning is different from afternoon. The question arises as to why one must include the frequencies just surrounding those two, at 364,365 and 729,731 cycles per year, the so-called sidebands or beat frequencies—see the power spectrum in Figure 1, where spikes are evident at those frequencies. This example is for the town of Mildura, Australia, latitude $-34.22°$ for the year 2004.

The concept of beat frequencies, also called sidebands, is well known in signal processing. In the language of that discipline, you can have a carrier signal with frequency $\omega_c = 2\pi f_c$ that has its amplitude modulated by a signal at lower frequency $\omega_m = 2\pi f_m$. The manifestation of this change in amplitude resides in signals at frequencies $2\pi(f_c \pm f_m)$, or in this case $2\pi\frac{(365-1)}{T}$ and $2\pi\frac{(365+1)}{T}$ for the daily cycle and a corresponding set of frequencies for the twice daily. T is the period, so $T = 8760$ for hourly data as an example.

Therefore, the Fourier series contains seven significant frequencies:

$$S_t = \alpha_0 + \alpha_1 \cdot \cos\frac{2\pi t}{8760} + \beta_1 \cdot \sin\frac{2\pi t}{8760} + \sum_{n=1}^{2}\sum_{m=-1}^{1}\left(\alpha_{nm} \cdot \cos\frac{2\pi(356n+m)t}{8760} + \beta_{nm} \cdot \sin\frac{2\pi(365n+m)t}{8760}\right) \quad (1)$$

The first article discussing the use of Fourier series, including the beat frequencies, as a means of identifying the seasonality of solar radiation data, was [23]. In it, Phillips argued that with the use of 75 Fourier coefficients, a 20 year data set of a climatic variable could be represented without significant loss of information. He used solar radiation as his test data, but gave an interesting example of an extension. He discussed the solution of a differential equation involving a mass of lumped thermal capacitance, exposed to a solar flux and losing heat to ambient temperature. If the Fourier transforms of the solar flux and ambient temperature have been calculated, the differential equation in the time domain is transformed to an algebraic equation in the frequency domain, affording a much easier approach to solution. This same approach was taken by the present author [24] to construct the analytic solution to the differential equations governing heat flows in domestic dwellings.

One of the reasons why one might choose the multiplicative modelling of solar radiation is the change of amplitude with level of the series. With solar radiation series and other climate data series, the amplitude of the daily cycle changes as one progresses through the year. The amplitude is higher in summer than winter, progressing systematically, rather than probabilistically, throughout the year. The Fourier series representation, by including the beat frequencies, captures this systematic amplitude modulation. It is a transparent and formal method of representing this modulation.

See Figures 2 and 3 for the effect of ignoring the sidebands. The model includes significantly non-zero values of solar radiation at night. Note that in these and subsequent figures, the term Data refers to the measured solar radiation values, and the term Model refers to the Fourier series representation of the data. Figure 4 illustrates the need for including the amplitude modulating frequencies. The data shown are the average daily values of solar radiation over the year, whereas the model is the Fourier series representation without the inclusion of the sidebands, averaged over the day. The values in the model at night have been zeroed as should be the case. This results in the bias shown with values too low in summer and too high in winter. Figure 5 depicts the same data as Figure 4, but the model now includes the sidebands frequencies. The addition of the sidebands means the model now follows the variation of the daily average in a more consistent manner over the year. It is internally consistent in that the physical interpretation of each term that is included is inherently simple and demonstrable. Figure 6 shows the performance of the Fourier series model with the sidebands included.

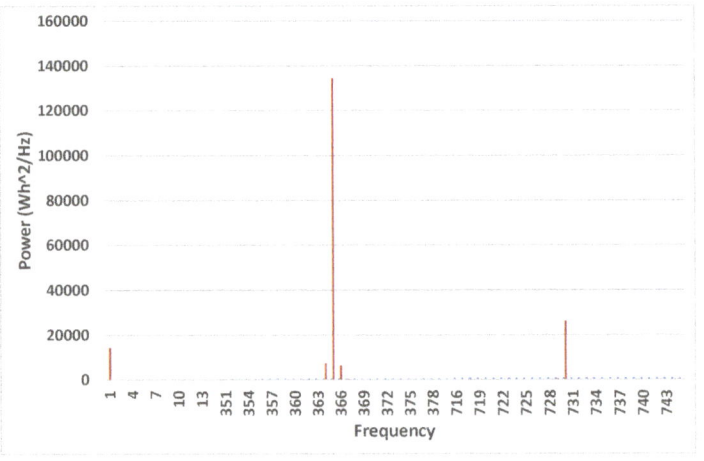

Figure 1. Power spectrum for hourly solar radiation for Mildura data.

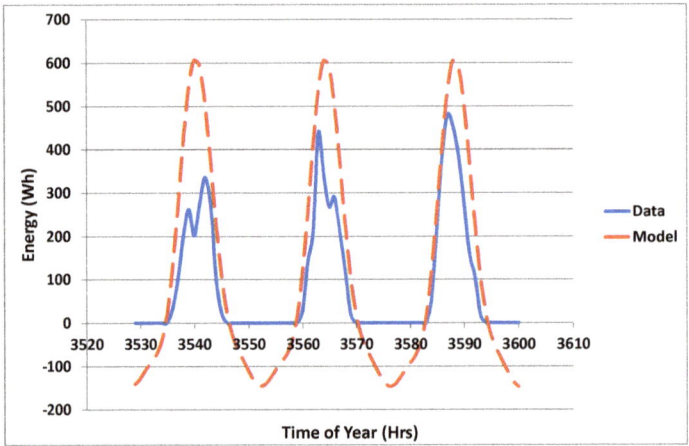

Figure 2. Effect of no sidebands on winter solar radiation.

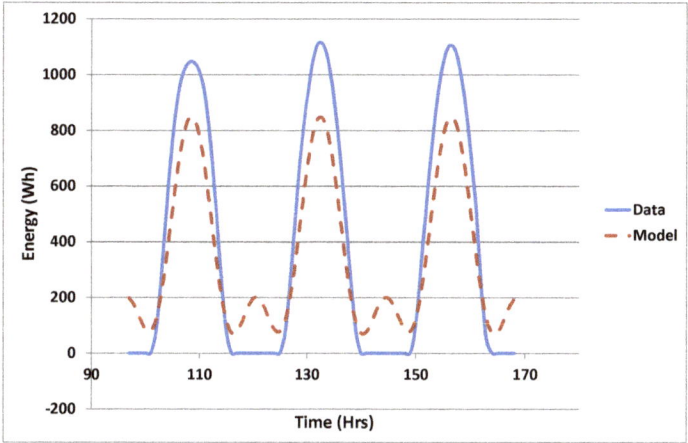

Figure 3. Effect of no sidebands on summer solar radiation.

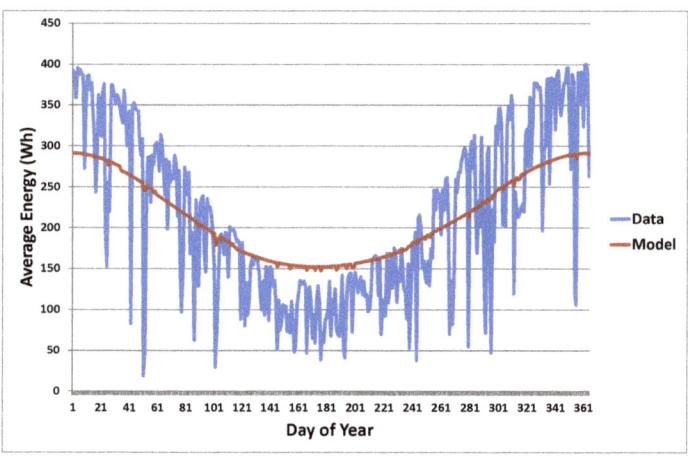

Figure 4. Daily Average Solar Radiation data plus model with no sidebands.

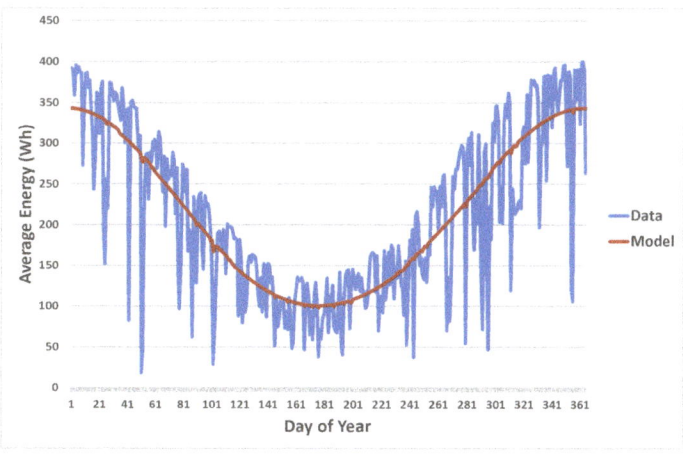

Figure 5. Daily Average Solar Radiation data plus model with sidebands included.

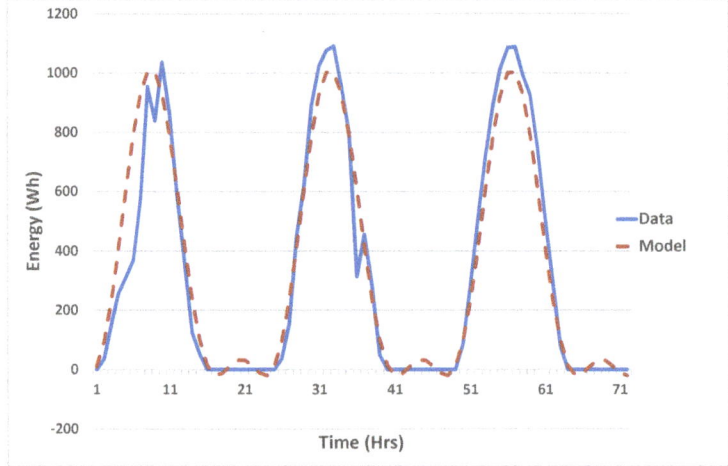

Figure 6. Effect of including sidebands on solar radiation.

3. Fourier Series Model for a Tropical Location

There is an interesting contrast in the analysis for a tropical location, the island of Desirade, part of Guadeloupe in the French West Indies, latitude 16.32°. Inspection of the power spectrum in Figure 7 hints at the fact that there may not be a significant change in the daily amplitude over the year. There are no apparent sidebands present in this graph. So, is this borne out? Let us examine a comparison of the data for a few days in summer plus a Fourier series model with sidebands and also one without the contribution from the sideband frequencies—Figure 8. There is very little difference with or without the contribution at the sideband frequencies. The difference between Desirade and Mildura can also be seen in Figure 9 for the daily mean solar radiation over the year for Desirade as compared with Figure 5 that shows the variation in daily amplitude over the year for Mildura.

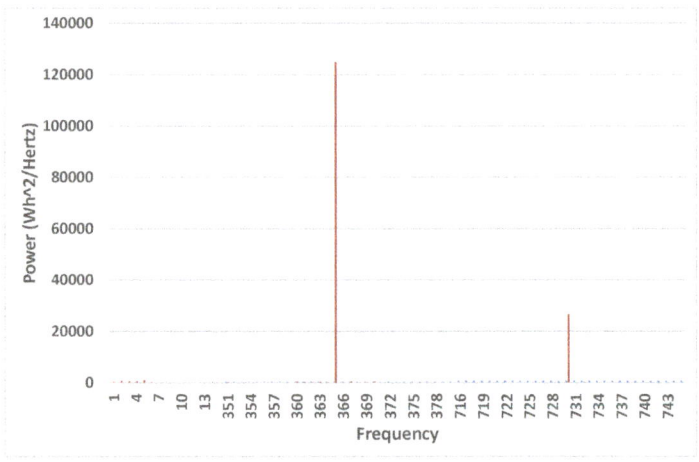

Figure 7. Power spectrum for Desirade.

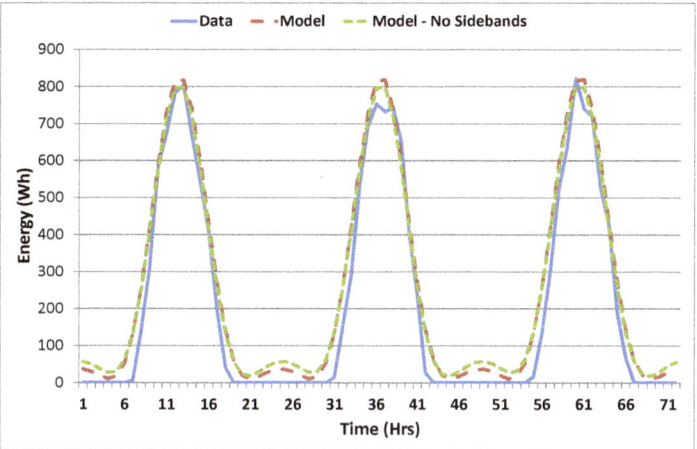

Figure 8. Comparison of Fourier series model with and without sidebands—summer in Desirade.

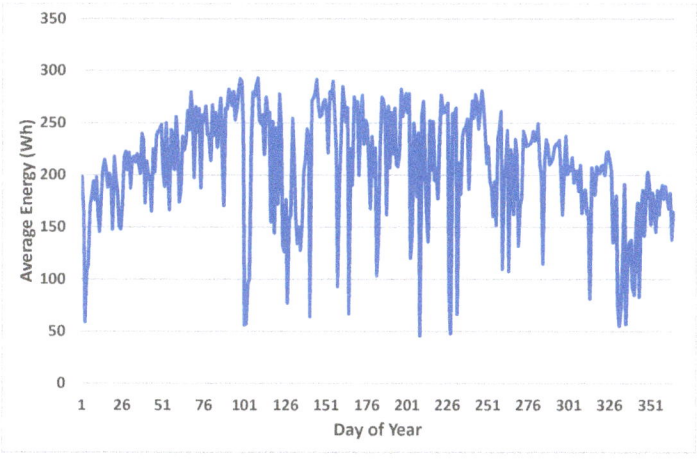

Figure 9. Daily mean solar radiation for Desirade.

An additional analysis was done for a location whose latitude is between Mildura and Desirade. St Pierre, Reunion Island, is at latitude −21.34°. An examination of the comparison of using the contribution from the sideband frequencies versus removing that contribution is given in Figure 10. It is obvious that the contribution at the sideband frequencies is more significant than at the tropical location, but less so than at the mid-latitude location, exactly as one might surmise. The decrease in importance of the sidebands for describing the seasonality as one traverses from mid-latitude in Mildura to lower latitude in St. Pierre, and even lower in Desirade, is due to the corresponding decrease in change of amplitude of the daily cycle over the year. As one draws closer to the equator, the daily amplitude approaches a constant. Interestingly, we will see a similar pattern with the output from solar farms in Australia in a subsequent section. Note that this type of information is not evident from using clear sky model multiplicative de-seasoning. Other limitations of the clear sky index approach will be given below.

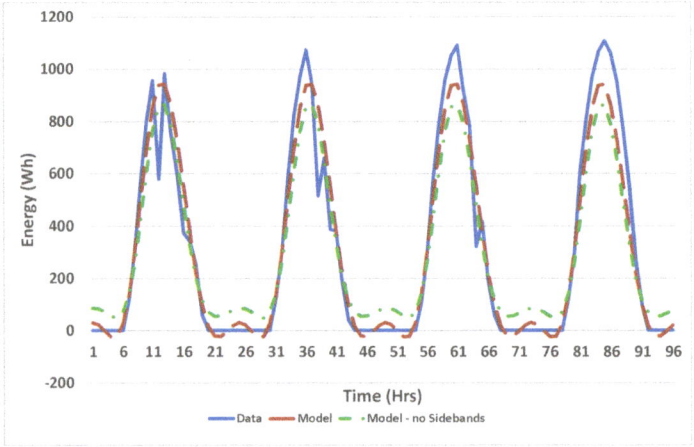

Figure 10. Comparison of Fourier series model with and without sidebands—summer in St Pierre.

4. Correspondence with Other Seasonal Climate Variables

There are situations where bivariate models of climate variables are necessary. If one of the components is solar radiation, then an alternative to the use of a clear sky model must be used, as there is no equivalent formulation for other climate variables. Examples have been given in [23,24]. Skeiker [21] illustrates that it represents the temperature seasonality very well. Even though there are situations where one does not have to treat all variables with the same methods, there is a good example of the need to do so with solar radiation and temperature. To model the performance of crystalline solar cells, it is necessary to build a bivariate model for the two variables. Ambient temperature has a lagged dependence on solar radiation, and as the efficiency of the solar cells is dependent on temperature as well as the incoming solar radiation, the two variables must be modelled in tandem. There is no temperature equivalent of a clear sky model, so an efficient method of modelling the seasonality of the two variables in a corresponding manner is through using Fourier series.

5. Clear Sky Models

The clear sky index (CSI) is defined as the solar irradiance divided by a suitable clear sky model. As stated previously, there are numerous clear sky models [20].

$$k^* = \frac{I}{I_{Clear}} \qquad (2)$$

One expects the clear sky index to be bounded in the interval $[0, 1]$. Because of the phenomenon of cloud enhancement (see [25]), it is possible to have some values greater than one. However, when we examine the CSI values for Las Vegas hourly data for 2010, constructed using the Bird clear sky model [26], we notice some problems. First, restrict the data to times for which the solar altitude $\alpha > 10°$, as there can be odd effects near sunrise and sunset. This stipulation is commonly used in evaluating forecasting for instance. Examining the histogram of values of CSI in Figure 11, the first problem is evident, as there are significant numbers of values greater than one. The second problem is visible in Figure 12, in that values at the beginning and end of each day for clear days are inflated. In essence one could say this is an induced seasonality, in that there is a U-shaped pattern from morning to afternoon.

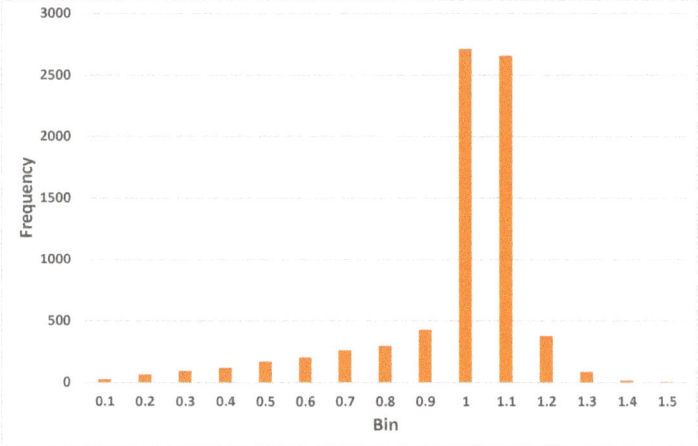

Figure 11. Histogram of CSI values for Los Vegas.

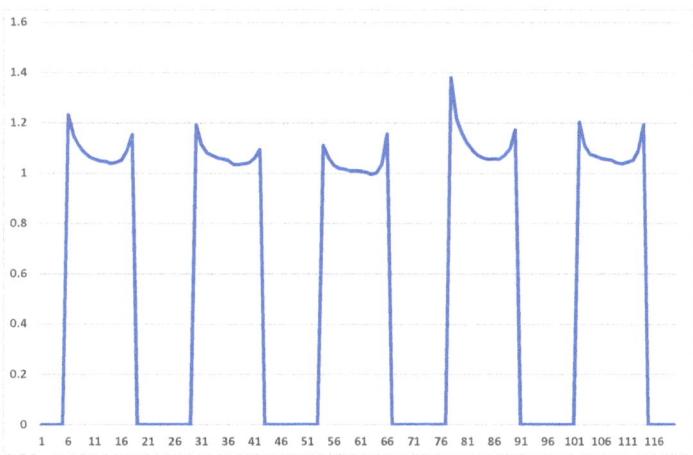

Figure 12. Five days of CSI for Los Vegas.

6. Solar Farms, the Australian Perspective

As of June 2019, Australia had over 12,000 MW of installed PV, with 4000 MW being commissioned in the previous 12 months. The largest solar farm is 220 MW (Bungala in SA) and there are thirteen over 100 MW. The Australian Renewable Energy Agency (ARENA) has funded a number of consortia to improve the very short-term forecasting of solar farm output. The present author, in conjunction with researchers from the University of New South Wales (UNSW) and the Commonwealth Science and Industrial Research Organisation (CSIRO) were awarded a grant in July 2019 for $1.25 million over 18 months for the project entitled **Solar Power Ensemble Forecaster**. The partner solar farms in the project are at Manildra, Clare, Darling Downs, Gannawarra, Parkes, Hayman, in the eastern states of Australia.

In Australia, the National Electricity Market (NEM) is controlled by the Australian Energy Market Operator (AEMO), who are also in charge of maintaining the electricity grid for the area that the NEM covers, Queensland, New South Wales, Victoria, Tasmania and South Australia. Because of its remoteness, Western Australia maintains a separate system. The NEM is unique in its operation in the following ways.

- Every 5 min, scheduled generators supply a bid stack, with the amount of energy they can provide in the next 5 min at each of 10 price bands, from −$1000 to $14,000.
- AEMO then runs a linear program for each region of the NEM to determine how far up the bid stack they have to go to satisfy their forecasted net load.
- This determines the 5 min price for all energy, and the mean of 6 five minute prices gives the spot price for the half hour.
- **Note**. There are also semi-scheduled and non-scheduled generators. Neither bid, but semi-scheduled can be curtailed in there is already sufficient supply in the system.

One big problem for AEMO is that their forecast model, the Australian Solar Energy Forecast System (ASEFS), is relatively crude. To the best of my knowledge they use a form of persistence, $\hat{S}_{t+1} = S_t$. Interestingly, the next few figures will show that the output is capped and so on a clear day the output is close to constant for a number of hours.

We begin by comparing the profile of solar radiation over the day in Figure 13 with that of solar farm output for both summer in Figure 14 and winter in Figure 15. For the radiation, it is for a clear day but for the output it can be for partially cloudy days as well. Obviously for the solar radiation on a clear day, there is a definite peak in the profile around solar noon, whereas for the farm output in both seasons, there is a definite cap. It is conjectured that this is for a specific reason, as solar panels have become relatively cheap in recent years compared to value of the electrical equipment for transferring the energy to the grid. Thus, it is relatively inexpensive to oversize the field. If, for instance, one has a power purchase agreement (PPA) with a customer, if one oversizes the field, it is easier to be confident of supplying the contracted energy on the majority of days. That lessens the need for purchasing energy on the spot market to supply the contracted amount.

This results in an interesting change in the power spectrum of the farm compared with that of solar radiation—see Figure 16. There is virtually no yearly cycle to the embedded in the data, and related to this, there are essentially no beat frequencies. This is consistent with output reaching capacity on days in both winter and summer. The power spectrum is similar to that of solar radiation for a location close to the equator like Desirade discussed above.

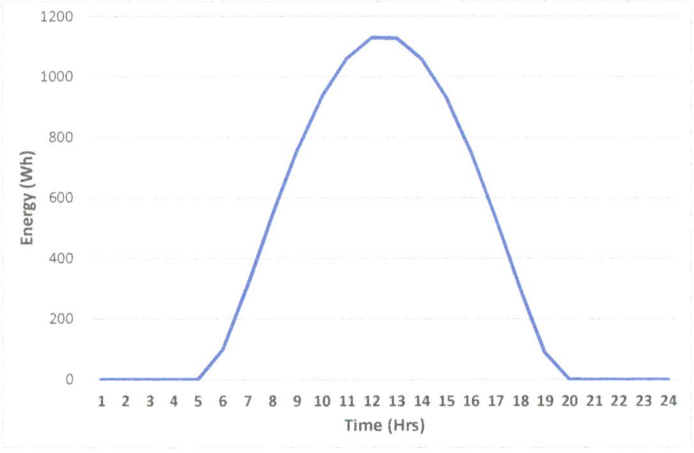

Figure 13. Solar radiation profile.

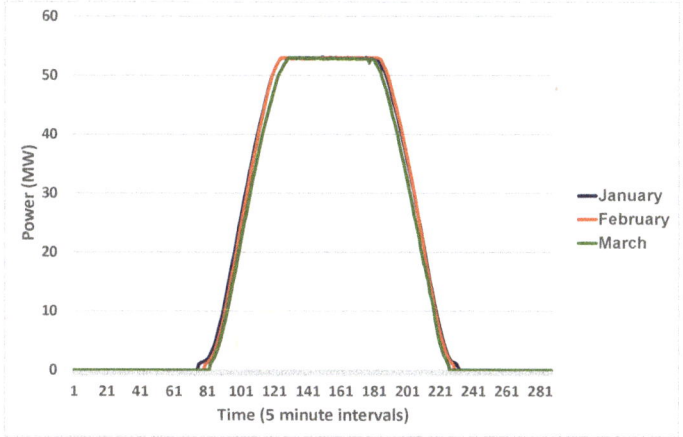

Figure 14. Clear day solar farm output, January–March.

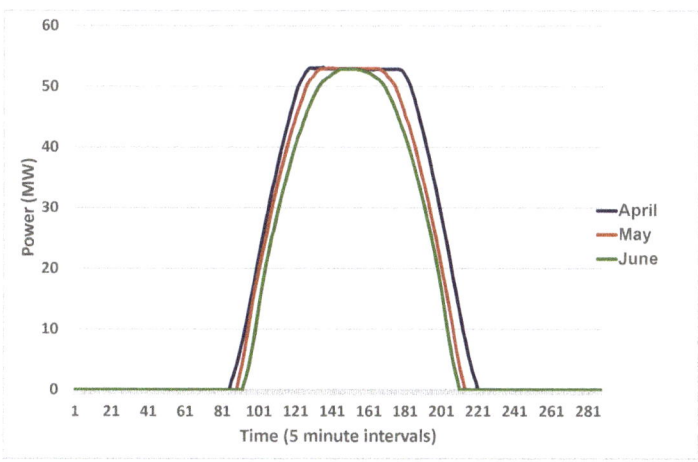

Figure 15. Clear day solar farm output, April–June.

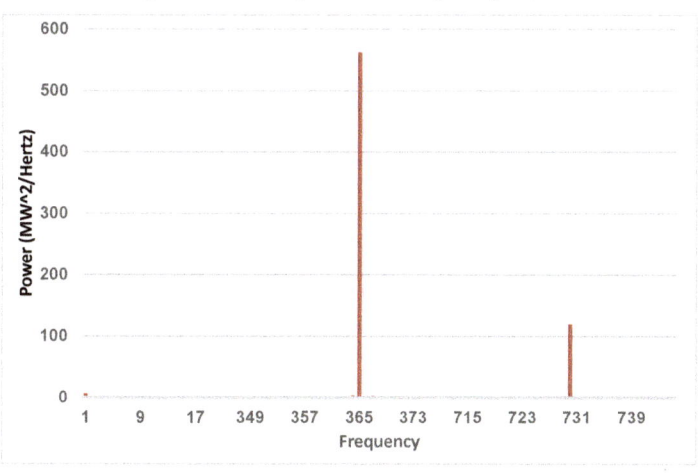

Figure 16. Solar farm output power spectrum.

7. Forecasting the Non-Seasonal Residuals

Any additive one step ahead statistical forecasting method can be encapsulated by the structure

$$Y_t = f(S_t; R_{t-1}, \ldots, R_{t-p}; X_{i,t-1}, \ldots, X_{i,t-q}) + Z_t \tag{3}$$

where $Y_(t)$ is the solar radiation, $R_t = Y_t - S_t$ is the difference between the solar radiation and the seasonality and S_t denotes the representation of the seasonality. The $X_{i,t}$ denote possible exogenous variables. Knowledge of the statistical qualities of the errors, or noise terms, Z_t is necessary in order to construct the error bounds of the forecast. In this formulation, it is hoped, and sometimes assumed, that Z_t is independent and identically distributed (i.i.d.).

For solar data, the Z_t are uncorrelated but dependent. Note that correlation is only a linear property, so higher moments can be, and are, correlated. The Z_t are not identically distributed—they vary both systematically, with the variance higher in the summer than winter and in the middle of the day compared to morning and afternoon. They also can vary time step to time step in a conditional manner. However, in what follows, we are not interested in forming error bounds on that forecast, so that will be left to future work.

After the seasonality model has been identified the algorithm for forecasting the de-seasoned data is as follows.

- Form $R_t = Y_t - S_t$, where Y_t is either solar radiation or the solar farm power output in MW, and S_t is the Fourier series representation of the climate.
- Check the sample autocorrelation function (SACF) and sample partial autocorrelation function (SPACF) to see what form of an $ARMA(p,q)$ should be used.
- There are two things to note here:
    - The Fourier series and $ARMA(p,q)$ models are estimated on a training set and then tested on a period of time not in the training set.
    - Many people at this stage use much more esoteric means for modelling, like ANN or other machine learning techniques.

Illustrative Results

In [12], the authors compare the use of Fourier series for seasonality and an autoregressive (AR) model (named CARDS) for forecasting the non-seasonal components with a number of models that combine clearness sky index and various ANN or ANN mixed with other tools. This is for forecasting solar radiation on an hourly time scale. The CARDS (coupled autoregressive and dynamical system) model performs at least as good as the other models. Since the forecasting of the non-seasonal component uses a basic low order AR model, the inference is that the Fourier series component is adept at handling the seasonality. Note that it was difficult to use exactly the same data as the researchers who developed the other models, but great care was taken to be as conservative as possible in setting up the experiments.

A more direct comparison was possible in [27], where the present author worked with colleagues from the Université de La Reunion to compare forecasting for island versus continental sites. A secondary goal was to compare the performance at both types of sites of the use of Fourier series plus an AR model (FSAR), clear sky index plus ANN and clear sky index plus AR. All three versions performed similarly in terms of the standard error measures of bias, mean absolute error and root mean square error. The salient difference is that the FSAR model is inherently simpler—and would be deemed so quantitatively if one used the Akaike or Bayesian Information Criteria for comparison. The components of the model all display knowledge of the climate of the site under consideration.

Note how the use of the Fourier series approach and an autoregressive model for the residuals once the seasonality is removed performs in an operational mode for the solar farm output forecast. Operational mode means that the forecast has to be made for a five minute interval at least 70 s before the beginning of the interval. This is to allow the communication of the forecast to AEMO so the mechanisms for any necessary frequency control or additional generation can be enacted. Also, the forecast mechanism is recalibrated, for both the Fourier series and autoregressive components, every five minutes in a rolling window. This is done to cater for any changing conditions in the farm or in the climate in the region around the farm. Preliminary comparisons were performed for four solar farms. This approach was found to outperform the method used at present by AEMO by between 8% and 36% over the four farms.

8. Conclusions

There are three methods in the literature for describing the seasonality of solar radiation, and all have been discussed here to lesser or greater degree. The clearness index formulation has been used but is probably more in use in the development of statistical models for diffuse solar radiation—see, for example, in [28]. For forecasting of solar radiation, the majority of practitioners would use the clear sky index. The crux of this paper is an argument for selecting the Fourier series method for describing the seasonality. The reasons include the following.

- There are several clear sky models so how does one choose the one to use? It may be that different ones work better in some climates and others in different climates.
- For the clear sky model described in this paper, there were technical difficulties in its application to data from Los Vegas.
- The components of the Fourier series representation have a direct physical interpretation.
- The Fourier series representation is compatible with the representation of seasonality of other climate variables, like temperature, and even some variables that are at least somewhat dependent on climate like electricity demand.

And finally, there is another important consideration. One can imagine that there was a very important practical consideration for adopting the use of the clear sky index. In Australia, for example, there are very few locations for which there are high frequency measurements of the components of solar radiation over an extended period of time. As the construction of the discrete Fourier series representation is empirically based, it is best to have at hand a number of years of reasonable quality data with which to estimate the coefficients, particularly if one is interested in hourly forecasting tools for instance. If instead, one can make use of a physical clear sky model, and then apply it to whatever short series is of interest to obtain clear sky index values, that might seem to be appealing. However, perhaps this argument is now superseded, as even where there are no ground measurements, there exist long periods of gridded data derived from models that estimate global horizontal radiation from satellite images. These are typically available for the hourly time scale and increasingly for higher frequencies. One can build Fourier series models from these data. Thus, there are compelling reasons for the use of Fourier series to represent the seasonality of both solar radiation and solar farm output.

Funding: This research was funded by the Australian Renewable Energy Agency (ARENA) grant for **Solar Power Ensemble Forecaster**—$1.25 million over 18 months. Chief investigators are Boland (UniSA), Kay (UNSW), Huang, West, Knight (CSIRO). The partner solar farms are Manildra, Clare, Darling Downs, Gannawarra, Parkes, Hayman.

Conflicts of Interest: The author declares no conflict of interest.

Abbreviations

The following abbreviations are used in this manuscript:

MDPI   Multidisciplinary Digital Publishing Institute
DOAJ   Directory of open access journals
TLA   Three letter acronym
LD   linear dichroism
AEMO   Australian Energy Market Operator
ARENA   Australian Renewable Energy Agency
UNSW   University of New South Wales
PPA   Power purchase agreement

References

1. Boland, J. Additive versus Multiplicative Seasonality in Solar Radiation Time Series. In Proceedings of the 21st International Congress on Modelling and Simulation, Modelling and Simulation Society of Australia and New Zealand, Gold Coast, Australia, 29 November–4 December 2015; pp. 1126–1132.
2. Inman, R.; Pedro, H.C.; Coimbra, C.M. Solar forecasting methods for renewable energy integration. *Prog. Energy Combust. Sci.* **2013**, *39*, 535–576. [CrossRef]
3. Diagne, M.; David, M.; Lauret, P.; Boland, J.; Schmutz, N. Review of solar irradiance forecasting methods and a proposition for small-scale insular grids. *Renew. Sustain. Energy Rev.* **2013**, *27*, 65–76. [CrossRef]
4. Sfetsos, A.; Coonick, A. Univariate and multivariate forecasting of hourly solar. *Sol. Energy* **2000**, *68*, 169–178. [CrossRef]
5. Kemmoku, Y.; Orita, S.; Nakagawa, S.; Sakakibara, T. Daily insolation forecasting using a multi-stage neural network. *Sol. Energy* **1999**, *66*, 193–199. [CrossRef]
6. González-Romera, E.; Jaramillo-Morán, M.A.; Carmona-Fernández, D. Monthly electric energy demand forecasting with neural networks and Fourier series. *Energy Convers. Manag.* **2008**, *49*, 3135–3142. [CrossRef]
7. Bacher, P.; Madsen, H.; Nielsen, H.A. Online short-term solar power forecasting. *Sol. Energy* **2009**, *83*, 1772–1783. [CrossRef]
8. Dong, Z.; Yang, D.; Reindl, T.; Walsh, W.M. Short-term solar irradiance forecasting using exponential smoothing state space model. *Energy* **2013**, *55*, 1104–1113. [CrossRef]
9. Cao, J.C.; Cao, S.H. Study of forecasting solar irradiance using neural networks with preprocessing sample data by wavelet analysis. *Energy* **2006**, *31*, 3435–3445. [CrossRef]
10. Cao, J.; Lin, X. Study of hourly and daily solar irradiation forecast using diagonal recurrent wavelet neural networks. *Energy Convers. Manag.* **2008**, *49*, 1396–1406. [CrossRef]
11. Mellit, A.; Benghanem, M.; Kalogirou, S.A. An adaptive wavelet-network model for forecasting daily total solar-radiation. *Appl. Energy* **2006**, *83*, 705–722. [CrossRef]
12. Huang, J.; Korolkiewicz, M.; Agrawal, M.; Boland, J. Forecasting solar radiation on an hourly time scale using a coupled autoregressive and dynamical system (CARDS) model. *Sol. Energy* **2013**, *87*, 136–149. [CrossRef]
13. Martín, L.; Zarzalejo, L.F.; Polo, J.; Navarro, A.; Marchante, R.; Cony, M. Prediction of global solar irradiance based on time series analysis: Application to solar thermal power plants energy production planning. *Sol. Energy* **2010**, *84*, 1772–1781. [CrossRef]
14. Aguiar, R.; Collares-Pereira, M. TAG: A time dependent, autoregressive, gaussian model for generating synthetic hourly radiation. *Sol. Energy* **1992**, *49*, 167–174. [CrossRef]
15. Cena, V.; Mustacchi, C.; Rocchi, M. Stochastic simulation of hourly global radiation sequences. *Energy* **1979**, *23*, 47–51.
16. Skartveit, A.; Olseth, J. The probability density and autocorrelation of short-term global and beam irradiance. *Sol. Energy* **1992**, *49*, 477–487. [CrossRef]
17. Lim, C.; Mcaleer, M. Forecasting tourist arrivals. *Ann. Tour. Res.* **2001**, *28*, 965–977. [CrossRef]
18. Reikard, G. Predicting solar radiation at high resolutions: A comparison of time series forecasts. *Sol. Energy* **2009**, *83*, 342–349. [CrossRef]

19. Paoli, C.; Voyant, C.; Muselli, M.; Nivet, M. Forecasting of preprocessed daily solar radiation time series using neural networks. *Sol. Energy* **2010**, *84*, 2146–2160. [CrossRef]
20. Ineichen, P. Comparison of eight clear sky broadband models against 16 independent data banks. *Sol. Energy* **2006**, *80*, 468–478. [CrossRef]
21. Skeiker, K. Mathematical representation of a few chosen weather parameters of the capital zone Damascus in Syria. *Renew. Energy* **2006**, *31*, 1431–1453. [CrossRef]
22. Boland, J. Time series and statistical modelling of solar radiation. In *Recent Advances in Solar Radiation Modelling*; Badescu, V., Ed.; Springer: Berlin/Heidelberg, Germany, 2008; Chapter 11, pp. 283–312.
23. Ruppert, M.G.; Harcombe, D.M.; Ragazzon, M.R.P.; Moheimani, S.O.R.; Fleming, A.J. A review of demodulation techniques for amplitude-modulation atomic force microscopy. *Beilstein J. Nanotechnol.* **2017**, *8*, 1407–1426. [CrossRef] [PubMed]
24. Boland, J. The Analytic Solution of the Differential Equations Describing Heat Flow in Houses. *Build. Environ.* **2002**, *37*, 1027–1035. [CrossRef]
25. Starke, A.R.; Lemos, L.F.L.; Boland, J.; Cardemil, J.M.; Colle, S. Resolution of the Cloud Enhancement Problem for One-Minute Diffuse Radiation Prediction. *Renew. Energy* **2018**, *125*, 472–484 [CrossRef]
26. Bird, R.E.; Hulstrom, R.L. *A Simplified Clear Sky Model for Direct and Diffuse Insolation on Horizontal Surfaces*; SERI–TR-642-761; Solar Energy Research Institute: Golden, CO, USA, 1981.
27. Boland, J.; David, M.; Lauret, P. Short term solar radiation forecasting: Island versus continental sites. *Energy* **2016**, *113*, 186–192. [CrossRef]
28. Ridley, B.; Boland, J.; Lauret, P. Modelling of diffuse solar fraction with multiple predictors. *Renew. Energy* **2010**, *35*, 478–483. [CrossRef]

© 2020 by the author. Licensee MDPI, Basel, Switzerland. This article is an open access article distributed under the terms and conditions of the Creative Commons Attribution (CC BY) license (http://creativecommons.org/licenses/by/4.0/).

MDPI
St. Alban-Anlage 66
4052 Basel
Switzerland
Tel. +41 61 683 77 34
Fax +41 61 302 89 18
www.mdpi.com

Energies Editorial Office
E-mail: energies@mdpi.com
www.mdpi.com/journal/energies

www.ingramcontent.com/pod-product-compliance
Lightning Source LLC
LaVergne TN
LVHW071957080526
838202LV00064B/6773